Seismic Mountings
for Vibration Isolation

SEISMIC MOUNTINGS FOR VIBRATION ISOLATION

JOSEPH A. MACINANTE

A Wiley-Interscience Publication

JOHN WILEY & SONS

New York   Chichester   Brisbane   Toronto   Singapore

Library of Congress Cataloging in Publication Data:

Macinante, Joseph A.
  Seismic mountings for vibration isolation.

  "A Wiley-Interscience publication."
  Includes index.
    1. Machinery—Vibration.   2. Damping (Mechanics)
I. Title

TJ177.M33 1984 620.3'7 83-21860
ISBN 0-471-87084-6

Printed in the United States of America

10 9 8 7 6 5 4 3 2 1

Foreword

Even though there has been widespread awareness of vibration and the problems it can cause for centuries, only within the last 50 years have college courses on the subject become relatively common. From the time the first formal texts on vibration engineering by Timoshenko and DenHartog were published in the 1930s to the present, developments in physics, techniques, and hardware associated with vibration have progressed at a steady pace. During these years published work in vibration isolation and damping—that of Crede, Henderson, Morrow, Snowdon, and Ungar, for example—also provided impetus for growth of the newly established field.

As has been typical of other technical areas, however, vibration control engineering has developed as needs and motivations have arisen. Principal motivations for vibration control today include the environments created by machines that rotate at relatively high speeds; the construction of lightweight machines, vehicles, and structures; and the performance sensitivity of electronic and optical equipment. During the period between 1940 and 1980 the average speed of rotating machines doubled—from 1800 RPM to 3600 RPM. But the forces of mass imbalance quadrupled; as a result much more vibration energy had to be dissipated or managed. The need to reduce construction costs stimulated the development of lightweight, flexible equipment. Such equipment has more resonant frequencies than heavier equipment, and these vibrations must be controlled. Equipment whose performance is highly sensitive to vibration environments is being increasingly used in medical, industrial, and military settings. Vibration control has thus become a challenging necessity for many engineers today.

In many cases control and suppression of vibration were and still are approached at the source. Often, however, the source cannot be altered to produce an acceptable environment for sensitive hardware; the electron microscope is a good example. Engineers have therefore had to approach environmental vibration problems by using secondary suppression; that is, isolation and damping. But isolation and damping create trade-offs among equipment, suppression hardware, and environment.

Despite an abundance of published material on vibration isolation and damping, they are among the least understood concepts in engineering. Myths abound on the power of damping in vibration attenuation. Too few engineers realize that, in most cases, additional damping increases transmitted vibratory forces because isolation and damping hardware introduce new natural frequencies into a system. And the difficulties of and limitations involved in converting vibratory energy into heat (damping) are too infrequently recognized. In fact too few engineers even know the difference between isolation and damping or are aware of the limitations of damping treatments and isolation hardware. Furthermore, perhaps because the theory is deceptively simple, the difficulties associated with properly applying isolation and damping technology are largely ignored. Another reason might be that current textbooks approach the subject from a theoretical point of view. *Seismic Mountings for Vibration Isolation* is concerned with the engineering *practice* of vibration control.

Even though Macinante has avoided the temptation to exhaustively cover the subject by referencing its theoretical and mathematical foundations, the work is scientific and rigorous. He emphasizes the physical and practical nature of vibration control engineering. Thus, most of the book is new and original. It will serve as a practical guide to vibration control engineering as well as a complete source of references in the field.

This well-written book will not only capture the interest of the reader but will also provide a wealth of valuable information. Each chapter is introduced by a quotation that Macinante skillfully incorporates into the subject. The material is organized so that the book can be used either as a text or as a handbook. In addition, it provides a complete survey of the isolation and damping area. The many design charts, case histories, and examples add immensely to its value to practicing engineers and architects.

Macinante is to be congratulated for this fine book. It will provide a boost to those engineers who must bridge the gap between theory and practice. And it puts to rest many of the misunderstandings and myths that have plagued the subjects of isolation and damping.

Ronald L. Eshleman

Vibration Institute
Clarendon Hills, Illinois
February 1984

Preface

Vibration can be pleasant and useful, and it can also be annoying and harmful. In this book we are concerned with the control of vibration of the latter kind, which causes trouble in various ways. In engineering works and factories the operation of engines, compressors, hammers, and other machinery generates vibration which is transmitted to surrounding areas and disturbs machine tools and other vibration-sensitive equipment. In office and residential buildings, vibration from a plant room commonly annoys the occupants of adjacent floors. In research and testing laboratories, vibration from air conditioning and workshop machinery may disturb electron microscopes and other sensitive apparatus. Sometimes the vibration is severe enough to damage equipment and buildings.

The ill effects of vibration can be eliminated, or reduced to acceptable levels, by effective vibration isolation, which should be planned and provided as part of the original construction of a building and its installations, not neglected and subsequently attempted by palliative action in response to complaints. It follows that architects and engineers should give serious attention to the vibration factor in the design of buildings, and in the layout and installation of the machinery and equipment that will be used in them. My aim in this book is to stimulate a concern for vibration control from an early stage in the planning and design of buildings, and to present a more realistic basis than is commonly used in practice for the design of the necessary vibration-isolating (seismic) mountings for machinery and sensitive equipment. The book is concerned with the concepts, not the "nuts and bolts" of the design process.

After more than 30 years of interest and involvement in vibration-isolation problems, it seems to me that in general practice, overseas as well as in Australia, seismic mountings are still being designed on the assumption that all will be well if the natural frequency of the mounting is appreciably lower than the frequency of the vibration that is to be isolated. A mounting designed on this basis for a vibrating machine may prove useless if the response of the supporting floor is ignored. A mounting for sensitive equipment may be ineffective if the response of the critical part of the equipment is ignored. My main objective in this book is to show how the flexibility of the floor in the one case, and that of the critical element in the other, can be taken into account by using a two-mass (two degree-of-freedom) design model instead of the one-mass model commonly used.

In general practice there seems to be some reluctance to consider the natural frequencies of mountings in modes other than the vertical. In 1951 Charles Crede showed how the six natural frequencies of a four-isolator symmetrical mounting can be evaluated by using elementary formulas for the two uncoupled modes and direct-reading design data for the other four modes. In this book I show how Crede's data can be applied also to a mounting on any number of isolators in any layout, not necessarily symmetrical, which satisfies certain simple conditions that minimize the coupling of modes and involve no difficulty in practice.

To the theorist the design basis offered in this book may seem oversimplified. Modern vibration theory used in conjunction with high-speed digital computers can determine the dynamic behavior of systems having literally thousands of degrees of freedom. The fact that these advanced techniques and facilities are used in the design of the hardware in special applications, for example, the NASA Space Shuttle, serves to highlight the gap between the available theory and that applied in the routine design of seismic mountings.

In pursuing these objectives I have tried to describe the physical basis of vibration isolation, and the application of new design data, in simple language and with a minimum of mathematics. There are many papers and books presenting the theory of vibration isolation in abstract and generalized mathematical language. At the other extreme, oversimplified versions appear from time to time in technical and trade publications. This book deals with the design of mountings in a way that is consistent with rigorous vibration theory, yet uncluttered with mathematical derivations and proofs that can be found elsewhere. The omission of mathematical material has been possible because the validation of the methods presented is published elsewhere, and references are cited for those interested in the details. It has not been necessary to include even the final expressions on which the design data are based, because they are incorporated in design data in the form of two- and three-dimensional computer plots which are more meaningful and convenient for use in the design office.

The major part of the book, Chapters 5 through 9, is concerned with the design of seismic mountings. The earlier chapters give the necessary background and terminology and, under the headings of vibration "insurance" and vibration criteria, offer guidance to those who must decide in particular circumstances whether a seismic mounting is necessary. Also, the earlier chapters include discussion of aspects of vibration measurement and dynamic balancing that relate to vibration isolation. The book is relevant to noise control only insofar as the provision of seismic mountings for machinery reduces structure-borne vibration and hence reduces the energy available elsewhere in the building for radiation as sound.

Throughout the book the emphasis is on the practical application of relevant and meaningful theory. Numerical examples, and references to the literature, are biased accordingly. Examples are given of good and bad practice, the latter serving to show the reality and seriousness of vibration problems that have occurred and highlighting the lessons that can be learned from the mistakes and omissions of others. In these examples from practice, the installation concerned is not identified, except in cases where a reference is cited and the installation is identified in the cited publication.

Although the book is addressed primarily to practicing engineers and architects, it should provide useful background reading for students in various branches of engineering and architecture, and for those engaged in special courses in vibration and noise control. Vibration consultants also may find it worthwhile to have a copy on the shelf for reference to the later chapters on the design of seismic mountings.

JOSEPH A. MACINANTE

Lindfield, Australia
February 1984

Acknowledgments

The subject matter of this book is based on the results of my investigation of vibration-isolation problems for the Commonwealth Scientific and Industrial Research Organization, Australia, Division of Applied Physics. I wish to express my appreciation of the publication policy of CSIRO which encourages the staff to write books so that the results of their work may be made available to as wide an audience as possible. In preparing the manuscript following my retirement in 1978, I have had the valuable assistance of the staff of the Division, particularly typists, librarians, and staff in the drawing office, photo laboratory, and administrative section.

In the experimental investigations and the computations referred to in the book, I have had the support of colleagues in the vibration group of the Division, whose specific contributions are acknowledged in the relevant contexts. For reading and offering constructive comments on the manuscript I wish to thank John L. Goldberg, who read the entire manuscript, Barrie Dorien-Brown for Chapters 5, 7, and 9, Norman Clark for Chapter 2, and David Eden for Chapter 8.

I would like to acknowledge the use of copyrighted material from the following sources:

The Opening Quotations:

Chapter 1: From *The World of Sound* by Sir William Bragg, G. Bell & Sons Ltd., London, 1933, p. 1, courtesy of Bell & Hyman Ltd.
Chapter 2: From W. J. M. Rankine, "The Mathematician in Love," in *A Book of Science Verse*, 1961, p. 150, selected by W. Eastwood, with permission from Macmillan & Co., London and Basingstoke.
Chapter 3: From *Shock and Vibration Digest* 1973, **5**(2), p. 1, courtesy of R. L. Eshleman.

Chapter 4: From Sven-Åke Axelsson, *Analysis of Vibrations in Power Saws*, Studia Forestalia Suecica No. 59, p. 9, 1968, courtesy of Royal College of Forestry, Sweden.

Chapter 5: From Alec B. Eason, *The Prevention of Vibration and Noise*, Oxford University Press, Oxford, 1923, p. 58, by permission of Oxford University Press.

Chapter 6: From R. E. D. Bishop, *Vibration*, 2nd ed., Cambridge University Press, Cambridge, 1979, p. 162, by permission of Cambridge University Press.

Chapter 7: from C. E. Inglis, "Mechanical Vibrations: Their Cause and Prevention," *Institution of Civil Engineers Journal*, 1944, pp. 316 and 318, courtesy of the Institution of Civil Engineers, London.

Chapter 8: From R. J. Steffens, "Some Aspects of Structural Vibration," in *Vibration in Civil Engineering*, Butterworths, London, 1966, p. 19.

Chapter 9: From T. N. Whitehead, *The Design and Use of Instruments and Accurate Mechanism*, Macmillan, New York, 1934, p. 184.

Other Quotations:

p. 114: From D. D. Barkan, *Dynamics of Bases and Foundations*, 1962, p. xi, reproduced with permission of McGraw-Hill Book Company.

p. 179: From T. V. Galambos, P. L. Gould, M. K. Ravindra, H. Suryoutomo, and R. A. Crist, "Structural Deflections. A Literature and State-of-the-Art Survey," U.S. National Bureau of Standards, Building Science Series 47, 1973, p. 2.

p. 179: From R. A. Crist and J. R. Shaver, "Deflection Performance Criteria for Floors," U.S. National Bureau of Standards, NBS Technical Note 900, 1976, pp. 1, 2.

p. 182: From *ASHRAE Handbook and Product Directory 1980 Systems*, pp. 35.22, 35.23, with permission of American Society of Heating, Refrigerating and Air-Conditioning Engineers, Inc.

Illustrations:

Figures 2.13 and 2.14: From J. L. Goldberg and P. Drew, "The Response of High-Rise Buildings to Ground Vibration from Blasting—an Experimental Investigation," 10th International Congress on Acoustics, 1980, Figs. 10(a) and 11.

Figures 2.15 and 2.16: From J. L. Goldberg and B. Dorien-Brown, "Image Motion in the Culgoora Solar Magnetograph—The Role of Vibration," Publications of the Astronomical Society of the Pacific, **84**(500), 1972, p. 537, Fig. 3, and p. 539, Fig. 6, courtesy of the Publications of the Astronomical Society of the Pacific.

Figures 3.2 and 3.3: From J. L. Goldberg, N. H. Clark, and B. H. Meldrum, "An Application of Tuned Mass Dampers to the Suppression of Severe Vibration of the Roof of an Aircraft Engine Test Cell," *Shock and Vibration Bulletin* **50**, Part 4, 1980, p. 61, Fig. 3, and p. 67, Fig. 13.

Figure 4.3: Adapted from "Balance Quality of Rotating Rigid Bodies," International Organization for Standardization, ISO 1940–1973, p. 8, Fig. 4, with permission through Standards Association of Australia.

Figure 5.6: From J. A. Macinante, "Spring Mounting for a Large Camera," *The Engineer*, **212**, 1961, p. 1081, Fig. 2, courtesy of Morgan-Grampian (Publishers) Ltd., London.

Figure 7.16: Adapted from C. E. Crede, *Vibration and Shock Isolation*, John Wiley & Sons, 1951, p. 54, Fig. 2.8.

Most of the illustrations and design data in Chapters 8 and 9 are reproduced from papers by the author with co-author Harry Simmons or Jeremy Walter, published by the Institution of Engineers, Australia.

A digest of the two papers on which the design data in Chapter 9 are based was published in the *Shock and Vibration Digest*, **8**(7), 1976, pp. 3–24.

Contents

Seismic Mountings
for Vibration Isolation

1

Introduction

All around us are material objects of many kinds, and it is quite difficult to move without shaking some of them more or less. If we walk about on the floor, it quivers a little under the fall of our feet; if we put down a cup on the table, we cannot avoid giving a small vibration to the table and the cup. If an animal walks in the forest, it must often shake the leaves or the twigs or the grass, and unless it walks softly with padded feet it shakes the ground. The motions may be very minute, far too small to see, but they are there nevertheless.

WILLIAM BRAGG (1933, p. 1)

In everyday usage the word vibration and others including oscillation, pulsation, shake, tremor, flutter, quiver—and even "wagging" as that of a dog's tail—denote to-and-fro motions of one kind or another. Vibrations manifest themselves in various ways. We see the vibration of a fluttering leaf or a twanged string. We feel the vibration of a springy floor when we walk on it, and the shaking of a vehicle when we travel on a bumpy road. We feel but seldom think about the vibrations that keep us alive—the beating of the heart and the motion of the diaphragm which pumps the air we breathe. Some vibrations are so small that we can detect them only with sensitive instruments.

Some vibrations are audible. Perhaps the most important are those of the vocal chords which generate speech and song. From the earliest times humans have been using vibrations to make music by plucking stretched strings, blowing on reeds or through tubes, and striking drums and gongs. Birds and insects use wing vibrations for communication as well as for flying. Cicadas, grasshoppers, and crickets make use of mechanical vibrations of particular parts of their bodies to generate song. Bragg's little book from which the opening quotation is taken is an enjoyable account of the ways in which vibrations become audible. ·

The present book is concerned only with vibration, specifically the control of vibration that is objectionable because it produces unwanted noise or other ill effects. This first chapter places our subject area in perspective. We begin by surveying the wide range of vibration that occurs in nature and as a result of human activities, and the ways in which we are aware of these vibrations. We note that vibration may be intentionally generated for useful purposes or may occur incidentally or accidentally. Whatever its origin, vibration may have objectionable effects on machinery, buildings, sensitive equipment, and humans. The need to eliminate these undesirable effects has stimulated the development of methods of vibration control that are dealt with in this book.

WIDE RANGE OF NATURAL AND MAN-MADE VIBRATION

Vibration and its effects are observed and studied in many contexts including those of molecular physics and chemistry; acoustics and seismology; and civil, electrical, mechanical, and aeronautical engineering. The vibrations of interest and importance are those of atoms and molecules; of air, water, and other fluids; of ground, buildings, and machinery; and of land, sea, air, and space vehicles.

The term *vibration* can be used to describe cyclical variation of any physical quantity, including pressure, stress, strain, electrical voltage, and current. In this book we are concerned only with the motion or change in position of vibrating objects.

A vibratory motion is commonly described in terms of two quantities: the *displacement amplitude*, which is the distance through which a vibrating point moves to one side or the other of its rest position, and the *frequency*, which is the number of to-and-fro cycles completed in unit time. The unit of frequency is the hertz, which is one cycle per second. Two vibrations having the same nominal amplitude and frequency can differ greatly in the detail of their motion, as illustrated in Chapter 2. In the following brief survey of the range of natural and man-made vibration, we refer simply to the displacement and the frequency.

Most of the effects of vibration on humans, machines, and structures are determined by the displacement in conjunction with the frequency. For a given displacement, the higher the frequency the greater must be the velocity and acceleration of the vibrating body in order to complete the greater number of cycles in a given time. For example, suppose it were possible to oscillate an entire building and its site horizontally through a distance of say 1 m (0.5 m amplitude). If the frequency were well below 1 cycle/min (1/60 Hz), there would be no noticeable effect on building or occupants, but if the frequency were 1 cycle/sec (1 Hz) the building would be wrecked.

The smallest vibrations are those of the atoms and molecules which make up all material things. These vibrations cannot be seen even with extremely powerful microscopes—their existence is postulated in scientific theory and demonstrated experimentally in the effects they produce. We take atomic vibrations for granted when we use electricity, and in our timepieces the atomic vibrations of crystals are replacing the mechanical oscillations that formerly controlled the mechanism. The atoms in a crystalline material are separated from one another by distances of the order of a few angstrom units (less than one millionth of a millimeter, see Fig. 1.1), and their vibrations take place in a small fraction of the interatomic spacing at frequencies as high as some tens of terahertz (10^{12} Hz).

The earth on which we live vibrates incessantly. Seismographs in operation all around the world for recording earthquakes show that even in the intervals between earthquakes the ground is always in a state of tremor. When a large bell is struck it goes on oscillating for many seconds; when our earth is shaken by a major earthquake it goes on oscillating for weeks or months. Seismologists and geophysicists detect and analyze these global oscillations and draw conclusions about the structure of the earth.

Microseisms or minor tremors of the earth's crust result from disturbances within the earth or on its surface. Storms at sea and the pounding of waves produce microseisms that can be detected hundreds of kilometers away. The ground displacement associated with these microseisms is mostly less than 0.01 mm and the frequencies below 1 Hz. Microseisms result also from industrial activities involving the use of heavy machinery and explosives, which produce microseisms that can be detected many kilometers from their sources.

The naked eye can see a vibration of some tenths of a millimeter, for example, the vibration of an instrument pointer. Vibrations are most easily seen when the frequency is higher than about 10 Hz in which range persistence of vision causes a vibrating point to be seen as a line.

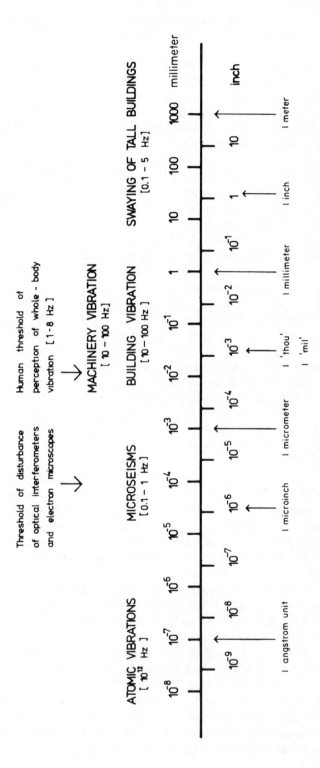

Figure 1.1. Range of natural and man-made vibration.

4

We may become aware of vibration through its effects on our bodies. As babies we were soothed by gentle oscillations in our mothers' arms and in our cradles. As children we enjoyed longer, slower oscillations on swings, and as we grow older we feel more secure and less conspicuous when we indulge in these relaxing oscillations in our rocking chairs. Vibratory devices have been designed into cushions and mattresses to impart beneficial vibrations to humans by relaxing muscles, increasing local blood circulation, and enhancing and lengthening periods of sleep—which unfortunately all too often is interrupted by the noisy vibration of an alarm clock.

Not all vibrations are pleasant or beneficial. We are sometimes subjected to annoying jolting and bouncing in buses and trains. We may suffer motion sickness in vessels in rough seas and in aircraft in turbulence. Nowadays it is not uncommon for the occupants of the uppermost floors in a very tall building to feel motion sickness when the building is buffeted by gusty winds. A less distressing but annoying vibration can be experienced in office and residential buildings where vibration from the air-conditioning plant can distract and annoy the occupants of adjacent areas.

In industry, machinery vibration can annoy the machine operators and others in the vicinity. The users of pneumatic rock drills and hand-held power tools for grinding, chipping, and rivetting can suffer disabilities of hands and fingers. The vibration and jolting experienced by the operators of tractors and earth-moving and mining machinery can cause spinal injuries. The personnel employed on off-shore drilling platforms are subjected to severe structural vibration and noise from the operation of machinery as well as from the action of sea and wind.

Humans are sensitive in one way or another to vibration in the frequency range from below 1 Hz up to 10 kHz. The magnitude of the vibration that humans can feel depends very much on the frequency and on the way the vibration is applied: through the feet, buttocks, head, hands, or fingertips. The sensitivity varies with the size, shape, and other physical characteristics of the person, as well as on psychological factors.

From the viewpoint of the comfort of a person subjected to vibration caused by machinery in buildings, the most important frequency range is 1–15 Hz, in which major resonances occur in the body. In the region of maximum sensitivity, 1–8 Hz, a person standing or lying can be aware of vibratory displacement amplitude smaller than 0.1 mm.

The range of vibratory displacement of machinery can extend from a fraction of a millimeter to one meter and more if we regard as vibration the motion of reciprocating parts of machinery such as the piston of a large engine, the worktable of a large planing machine, and the oscillating parts of machinery used for vibratory processes of the kinds mentioned below in the section Useful Vibration.

However, when we speak of machinery vibration we have in mind the vibration of shafts, bearings, frame, and baseplate that is incidental or unnecessary to the functioning of the machine, and which can serve as an indicator of the operating condition or "well-being" of the machine. Usually the predominant component of the vibration results from imperfect dynamic balance of the rotating parts, in which case the vibration frequency is equal to the shaft rotational frequency. A bearing displacement amplitude of 0.01 mm at 10 Hz is good, and 1 mm nonpermissible. At 100 Hz a displacement of 0.001 mm is good and 0.1 mm nonpermissible.

The vibration of the floors and walls of buildings in this frequency range of 10–100 Hz, resulting from the operation of machinery, is generally well below 0.1 mm. The magnitude of vibration that could cause damage to masonry walls and plaster ceilings depends, of course, on the type, age, and quality of the construction. On the basis of experience and tests, the damage threshold is represented by a displacement amplitude of about 0.1 mm at 10 Hz and 0.01 mm at 100 Hz.

A much higher range of building displacement is associated with the very low-frequency swaying of tall buildings under wind gusting. The displacement may be some hundreds of millimeters at frequencies below about 5 Hz.

USEFUL VIBRATION

This book is primarily concerned with the control of unwanted vibration. Lest it appear that vibration is an unmitigated evil, we now mention some of the ways in which vibration is intentionally generated for useful purposes. Unfortunately, a common by-product of most of the useful applications is some unwanted vibration.

Mechanical vibrators of many kinds are used in industry for operations such as mixing, stirring, and shaking which are necessary in processing materials and manufacturing products. Small objects can be moved on vibratory conveyors from one operation to another along an assembly line. In each cycle of the vibration the objects make a small step forward when the conveyor track moves upward and forward, and when the track moves in the opposite direction the objects do not slip back but make another step forward on the next cycle of the vibration.

Loads can be moved by vibration. In one device a skid carrying the load has a power-driven oscillating mass. The line of motion of the mass is inclined at an angle to the vertical. When the mass is at the upper end of its travel it applies a force that is upward and forward; this reduces friction at the skid, which slides forward. When the mass is at the lower end of its travel, the force is downward and backward, thereby increasing the friction and the skid does not move backward. The line of oscillation of the mass can be tilted to move the load forward, backward, and sidewise.

Manufactured components are cleaned by vibrating them while they are suspended in an abrasive medium, which cleans and deburrs otherwise inaccessible areas. Farm eggs are cleaned in a similar way for marketing.

Thermoplastic components are welded by pressing together the surfaces to be joined and vibrating them in the plane of the joint; the heat generated by friction melts the material, which remains joined when it solidifies again.

Metallic parts and assemblies become internally stressed as a result of shrinkage, tool cutting forces, and any other working to which they are subjected in the course of manufacture. If unrelieved, these stresses can cause distortion of the material and hence malfunction. For many years stress relief has been done by thermal methods; now it may be done by suitable vibration imparted to the object. The theory of stress relief by vibration is not well developed. It is based on the idea that distortion of the unstabilized material results from imperfections in the molecular crystal lattice, and that mechanical vibration of the material causes a movement of the imperfections in a manner that reduces the level of residual stress.

One of the most important uses of vibration in industry is for the vibration testing of products and assemblies that will be subjected to vibration in service. Vibration testing is most extensively used in the aircraft industry. Vibration is imparted to the prototype aircraft or spacecraft, or to a scale model or a particular part, and the response is observed to discover any malfunction, cracking, or other weakness. Any necessary design modifications are made to increase resistance to vibratory fatigue failure. Samples of the materials and of actual components used in the aircraft wings, fuselage, and engines are fatigue tested to determine the life expectancy under their working conditions of alternating stress and strain. Vibration testing is applied also in the investigation of the effects of sustained vibration on the efficiency and comfort of pilot and crew.

Knowledge of the elastic and internal damping properties of materials is important in engineering design for dynamic loading conditions. Vibratory test methods are used to acquire design data and to study theories on the structure and damping mechanism of metallic and elastomeric materials.

Vibration testing is also widely used in the packaging industry to ensure that the package can protect the contents from the impacts and vibration that the package will experience during transport from manufacturer to user.

In making roads and preparing building sites, vibrating plates and vibratory rollers are widely used for compacting materials ranging from crushed rocks to bituminous mixes. The vibration adds to the downward force imparted and can be adjusted to the optimum conditions for the particular type of material.

The traditional method of driving a pile is by striking it repeatedly with a heavy blow. In reconstruction work in built-up areas the vibration resulting from these impacts can damage existing structures by causing settlement of foundations and cracking of walls. This can now be avoided by using a vibratory method in which the pile, or the caisson in which it will be formed, is vibrated mechanically. The agitation of the layer of soil adjacent to the pile reduces friction between pile and soil, and the pile or caisson sinks slowly and quietly through the ground to rest on a bearing stratum.

Even the unwanted vibration of running machinery can be useful because it contains information about the well-being or operating health of the machine. A vibration specialist can diagnose vibration and thereby identify a malfunctioning or faulty component in much the same way that a heart specialist can interpret an electrocardiogram to identify faulty heart function.

Vibration analysis is applied also in seismic prospecting and site investigation. The nature of the vibration transmitted through the ground from an explosive charge gives information about the material under the surface between source and receiver. Techniques have been developed for the analysis of ground vibration from a distant explosion to determine whether or not the explosive was a nuclear device.

HARMFUL VIBRATION OF MACHINERY, INSTRUMENTS, AND STRUCTURES

Vibration is a common source of trouble in engineering works and factories where the operation of engines, compressors, hammers, presses, and other machinery generates vibration that is transmitted to surrounding areas and disturbs precision machine tools, measuring instruments, and sensitive manufacturing processes.

The harmful effects begin in the machinery that is the source of the vibration. Unbalanced forces generated by rotating and reciprocating parts cause the whole machine to vibrate on its supports and also give rise to internal stresses. The vibration may be severe enough to cause malfunctioning of the machine, excessive wear and damage to bearings, couplings, and other parts, fatigue failure of shafts, and even the destruction of the machine.

Sensitive equipment and processes are often, for reasons of production flow, located near sources of vibration. This may result in poor quality of work from fine machining operations, errors in the readings of instruments, and malfunctioning and damage of delicate apparatus. In hospitals and teaching and research establishments, vibration from air conditioning and other machinery can disturb sensitive optical, electronic, and other instruments.

It is obvious from the effects of earthquakes that vibration can damage buildings and other structures. Vibrations generated by engineering activities are less severe, yet can be damaging. For example, the use of explosives for such purposes as excavation, demolition, quarrying, and tunneling generates ground vibration, which can cause significant cracking of the walls of masonry buildings in the vicinity. However, in many instances the most harmful effect is disturbance of the peace of mind of the property owner. Vibration of large unbalanced engines and compressors can have similar effects. Structural damage can be caused directly by vibration of the structure itself or indirectly as a result of settlement caused by vibration of the supporting soil and foundations.

In the foregoing we have referred to damage that can be caused by sources of oscillatory energy inside or outside the structure. It is possible for a structure to be oscillated by a source of energy that is not oscillatory, in the same way that a steady wind can cause a leaf to vibrate. Such a self-induced or self-excited vibration results from the conversion of nonoscillatory energy into oscillatory excitation within the system itself. The periodic excitation caused by a steady wind results from the shedding of eddies as air flows past the structure. The best known and most spectacular example is the failure of the Tacoma Narrows suspension bridge in 1940. Other examples are the wind-induced "galloping" of electrical transmission lines and the vibration of tall masts and chimneys. A lucid nonmathematical discussion of these "vibrations that cause themselves to grow" is given by Bishop (1979, Chap. 4).

CONTROL OF UNWANTED VIBRATION

Clearly vibration can be pleasant and useful, and it can be unpleasant and harmful. Vibration can interfere with our comfort, working efficiency, and health, and can disturb and damage instruments, machinery, and structures. Even vibration that is imperceptible to humans can be detrimental in industrial, technological, and scientific activities.

It follows that architects and engineers should give serious attention to vibration control when designing buildings and planning the layout and installation of machinery and equipment. The aim of this book is to help them do so.

The terminology of vibratory motion and methods of vibration measurement are discussed in Chapter 2. The consideration of vibration control begins in Chapter 3, which contains a survey of the various lines of action that may be taken to solve existing vibration problems and to reduce the risk of trouble with proposed new installations. Among the possibilities is that of installing vibration sources and vibration-sensitive equipment on mountings designed to minimize vibration transmission. These are referred to as *seismic mountings* for the following reason. A seismic system is simply a mass attached to a base by means of springs. For example, in a seismograph a large mass is supported by a "soft" spring suspension, which is mounted on a base or frame that is set in the ground. The mass remains almost stationary when the ground vibrates; hence, the mass provides a reference or datum for the measurement of the ground vibration.

A seismic mounting is necessary if the vibration severity without the mounting would be likely to cause structural damage or disturb sensitive equipment or humans. Therefore, in the design of new buildings and installations it is necessary to estimate the levels of vibration that the machinery may cause in the plant room and in vibration-sensitive areas in the building, and to compare these levels with permissible or acceptable levels of vibration. To do this we need to know what vibration is acceptable; that is, we need *vibration criteria* for machinery, buildings, instruments, and humans. The vibration criteria currently available are discussed in Chapter 4.

SEISMIC MOUNTINGS

The latter and major part of the book, Chapters 5 through 9, is concerned with seismic mountings. A seismic mounting is formed by interposing resilient material, typically in the form of metal springs, rubber isolators, or air springs, between the equipment that is to be isolated and its support. The term "support" denotes the floor or other structure on which the seismically mounted equipment is installed.

In text and reference books we find terms such as "vibration-isolating system," "spring mounting system," "spring foundation," "spring suspension," and "antivibration mounting." In this book we prefer the term seismic mounting.

In a seismic mounting designed to isolate sensitive equipment from vibration of the support, the seismic mass, like that in a seismograph, does not move significantly when the support vibrates: the resilient material isolates the mass from vibration of the support. Alternatively, a seismic mounting can be designed for equipment that is a source of vibration to reduce the vibration transmitted from the machine to the support.

A mounting for a source of vibration is conveniently called a *source mounting*, but there is no concise term in general use for a mounting that is intended to isolate sensitive equipment from external vibration. The terms "positive" and "negative" isolation used by some writers for source and receiver isolation, respectively, are vague, and, outside the context where defined, meaningless. The terms "active" and "passive" isolation may be misleading because some mountings are called active because they are servo controlled and, in this sense, all other mountings, whether for source or receiver, are passive. In this book we use the term *mounting for sensitive equipment* or *receiver mounting* when the context is concerned with the relationship between source and receiver.

The hardware of seismic mountings is described in Chapter 5 in which the basic types of mounting are identified and the various kinds of resilient material used for vibration isolation are discussed, particularly helical springs, rubber isolators, and air springs. The design of a seismic mounting involves a number of practical as well as technical considerations. Chapter 5 gives some guidance on the important practical matters and on the specification and performance testing of seismic mountings.

In order to establish a working basis for the design of a mounting, we need to make some simplifying assumptions about the proposed installation and its response to vibration. These assumptions are embodied in a design model, which is a conceptual idealization of the installation. Chapter 6 shows the inadequacy of the one-mass model commonly used and argues the merits of the two-mass model adopted in this book.

A seismically mounted mass will vibrate freely if pushed then released. It can be made to bounce and rock in various modes of free or natural vibration, and it has a certain well-defined natural frequency of vibration in each of these modes. When vibration is forced or imposed on either the seismic mass or the support, the quality of vibration isolation achieved by the mounting is determined by the relationship of these natural frequencies to the frequency of the imposed vibration and to certain other frequencies, mentioned in the next paragraph, that are inherent in the installation. The design task is to decide what should be the natural frequencies of the mounting, and to choose the resilient material and arrange its configuration to achieve the desired natural frequencies. The physical nature of free vibrations and the method of calculating natural frequencies are explained in Chapter 7.

The factors that determine desirable values for the natural frequencies of a mounting for equipment that is a source of vibration are not the same as those involved when a mounting is required to isolate sensitive equipment from site vibration. Source mountings are discussed in Chapter 8, which shows how the flexibility and hence the natural frequency of the supporting floor or structure influence the choice of the natural frequencies of the mounting. Chapter 9 deals with mountings for sensitive equipment and shows how the natural frequencies of the responsive or critical elements of the equipment influence the choice of the natural frequencies of the mounting.

REFERENCES

Bishop, R. E. D. (1979). *Vibration*, 2nd ed. Cambridge University Press, Cambridge, England.
Bragg, William (1933). *The World of Sound*, G. Bell, London.

2

How to Describe Vibration

The lady loved dancing:—he therefore applied,
To the polka and waltz, an equation;
But when to rotate on his axis he tried,
His centre of gravity swayed to one side,
And he fell, by the earth's gravitation.

W. J. M. Rankine (1874)

It would be difficult to describe the combination of oscillation and rotation in which the dancer engaged before becoming unstable under the action of centrifugal and gravitational forces. We begin with the simpler kinds of oscillation.

Although the term "to-and-fro" may serve well in everyday usage, in technical contexts we define a vibration to be a cyclical change in the position of an object as it moves alternately to one side and the other of some reference or datum position. This is the definition of vibratory motion, which is our primary concern in this book, although the terms to be defined are applicable to any physical quantity.

The first part of this chapter introduces the terminology of the physical quantities that characterize vibration. The second part shows how vibration measurement assigns numerical values to these physical quantities. Vibratory motion may take various forms, which can be described conveniently only by diagrams or mathematical expressions. We begin with the vibration of a point whose path is a straight line; this is rectilinear or translational vibration. We then discuss rotational and combined translational and rotational vibration of a body as a whole, regarding it as a rigid body. Finally, we refer to flexural and other elastic vibrations or distortions of a body associated with its nonrigidity or elasticity.

This consideration of the physical nature of vibration is necessary background for the discussion of vibration isolation, because the characteristics of the vibrations involved determine the design of seismic mountings. Although practicing engineers and architects who require the measured data are not expected to have the experience or equipment to make the necessary vibration measurements, which nowadays is a matter for the specialist, they do need sufficient understanding of the principles and practice of vibration measurement to enable them to communicate effectively with the vibration specialist, to ensure that the measurements made are meaningful, adequate, and no more detailed than necessary.

TERMINOLOGY OF VIBRATORY MOTION

Since this chapter is addressed to those who have not previously taken much interest in the subject, no prior knowledge of vibration is assumed, and the scope is limited to terminology that is relevant to normal practice in vibration isolation. Comprehensive terminologies are given by the International Organization for Standardization (1975) and Harris and Crede (1976, Chap. 1, Appx. 1.2).

Sinusoidal Vibration

If we look at a point vibrating in a straight line, vertically for example, we can watch it go up and down if it moves slowly enough. If it vibrates rapidly the point will blur and be seen as a line. In order to describe in any detail the cyclical variation of the position of the point, we make use of a *waveform diagram*, which is a graph showing how the position or displacement of the point varies with time.

The most familiar form of vibration is simple harmonic motion, usually called *sinusoidal* or *harmonic*, which is illustrated in Fig. 2.1. The point P (Fig. 2.1a) is vibrating in the vertical (z) direction; its displacement waveform is a sine curve as shown in Fig. 2.1b. The symbol z is used here in anticipation of subsequent discussion of horizontal and vertical vibrations, and rotational vibration about horizontal and vertical axes (e.g., see Fig. 2.10), for which x and y are used to denote the horizontal axes and z the vertical.

If the point P is moving upward past the reference or datum position O at the time $t = 0$, then the displacement z of the point at any instant t seconds later is given by

$$z = Z \sin \omega t \tag{2.1}$$

where Z is the maximum or peak displacement of P from its datum position.

The reason for the intrusion of ω can be seen by referring to Fig. 2.1c. The motion of point P in Fig. 2.1a is identical with that of the point P' in Fig. 2.1c where P' is the foot of the perpendicular QP' drawn from a point Q that is rotating anticlockwise, with angular velocity ω radians per second (rad/s), in a circular path of radius Z. The angle ωt in Eq. (2.1) is the angle through which $O'Q$ turns in time t from its initial position $O'M$. The angular velocity is usually called the angular frequency (rad/s), and the number of cycles per second is the cyclical frequency called hertz (Hz). In practice, the word frequency is commonly used for both the angular and the cyclical frequency—the context indicates which is intended.

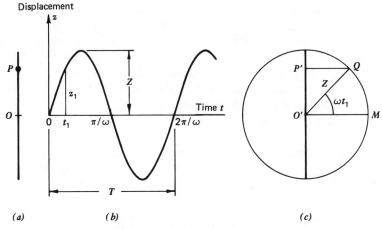

Figure 2.1. Sinusoidal vibration.

The time (T) taken for point P in Fig. 2.1a to make one complete cycle of vibration is equal to the time for the radius $O'Q$ in Fig. 2.1c to make one complete revolution, 2π rad. Therefore, $T = 2\pi/\omega$ and the number of cycles per second is $f = 1/T = \omega/2\pi$ Hz.

Since the vibrating point momentarily comes to rest at an end of its excursion, it must accelerate in order to move again and acquire velocity. Then it must decelerate to come to rest again momentarily at the other end of its travel, and so on. Thus we can, and when necessary do, describe the motion in terms of the cyclically varying velocity or acceleration, as illustrated later.

A decade or two ago the term amplitude was generally understood to mean the displacement amplitude (Z) and the term "double amplitude" was used to denote the total displacement ($2Z$). Today the term amplitude is not reserved for displacement but is used to denote also the peak value of velocity, acceleration, or any other oscillatory physical quantity. The term double amplitude is not used because it has meaning only for vibrations having equal positive and negative peak values; the term *peak-to-peak* displacement is preferred.

Physically, the instantaneous velocity is the rate of change of displacement, and the acceleration is the rate of change of velocity. Mathematically, in the terminology of the calculus, the velocity is the first time derivative (dz/dt) of the displacement, and the acceleration is the second time derivative (d^2z/dt^2). A convenient notation for the first and second derivatives uses one and two dots, respectively, over the symbol for the displacement, as in the following equations, which are derived in textbooks (e.g., Den Hartog, 1956, Chap. 1).

For sinusoidal motion:

$$
\begin{array}{lll}
\text{Displacement} & z = Z \sin \omega t \\
\text{Velocity} & \dot{z} = \omega Z \cos \omega t & \text{(2.2)} \\
\text{Acceleration} & \ddot{z} = -\omega^2 Z \sin \omega t
\end{array}
$$

It may be noted in these expressions that the acceleration is negative when the displacement is positive, and vice versa. If the acceleration were not of opposite sign to the displacement there could be no vibration: a negative acceleration is simply a retardation, and it is this retardation that brings the oscillating point to rest at the end of each excursion and accelerates it back toward the mid-position, as illustrated by the following example. In practice we are concerned with the amplitude or peak value of the acceleration, and the negative sign is ignored.

Example 2.1.   Evaluate the velocity and acceleration of the point P in Fig. 2.1, if the displacement amplitude is $Z = 10$ mm and the frequency is 5 Hz (period $T = 0.2$ s) as shown in Fig. 2.2a.

The cyclical changes in the velocity and acceleration associated with the sinusoidal displacement shown in Fig. 2.2a are given in Fig. 2.2b and 2.2c,

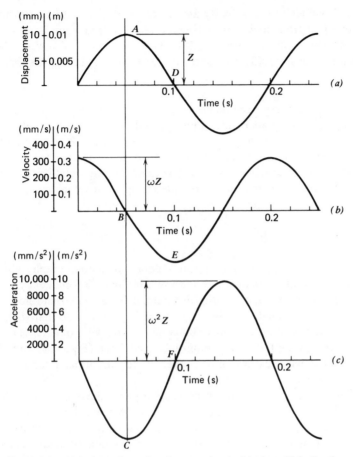

Figure 2.2.   Velocity and acceleration associated with sinusoidal vibration.

respectively. These waveforms are those of the expressions given in Eq. (2.2) for velocity and acceleration. It can be verified by inspection that the ordinate of the velocity curve at any instant is proportional to the rate of change (gradient) of the displacement curve, and that the ordinate of the acceleration curve at any instant represents the rate of change of the velocity.

Thus, when P is momentarily stationary (i.e., velocity = zero) at the top of its excursion (point A), the velocity shown in Fig. 2.2b is zero (point B). At this instant the acceleration (point C) has its maximum negative (downward) value which is consistent with the maximum negative gradient of the velocity curve at B. This acceleration causes the point to acquire downward velocity (along the curve BE) until, as P passes down through its mid-position (point D), the velocity has its peak negative value (point E). At this instant the acceleration is passing through zero (point F) and becoming positive, thereby retarding the downward movement, and so on.

The peak values of the velocity and acceleration are evaluated as follows. For the cyclical frequency $f = 5$ Hz, the angular frequency is $\omega = 2\pi f = 31.4$ rad/s. The peak velocity is $\omega Z = 31.4 \times 10 = 314$ mm/s. The peak acceleration is $\omega^2 Z = (31.4)^2 \times 10 = 9860$ mm/s^2 or approximately 9.9 m/s^2.

From the foregoing we see that during a sinusoidal motion with a particular frequency ω, the displacement, velocity, and acceleration are continuously changing within limiting or peak values of Z, ωZ, and $\omega^2 Z$, respectively. The motion can be defined in terms of any two of the quantities: frequency, displacement, velocity, and acceleration. For example, data relating to machinery and structural vibration are usually presented in terms of velocity and frequency, or acceleration and frequency.

Vibration Nomograph

A convenient way of presenting vibration data is on a nomograph, which may have various formats, one of which is shown in Fig. 2.3. Frequency is plotted as the abscissa and velocity as the ordinate. Displacement and acceleration are plotted on a diagonal grid. The scales are logarithmic so that wide ranges of numerical data can be shown. Usually several decades of each variable are shown as, for example, in a nomograph used later (Fig. 4.4), but for the present purpose we illustrate only about one decade of each variable so that detail can be shown to a more open scale.

Any point on the nomograph represents a particular sinusoidal vibration, whose peak value of displacement, velocity, and acceleration for a given frequency are indicated on the corresponding scales. Any two of these quantities may be used to define the point; then the other two quantities associated with that vibration can be read off the relevant scales. For example, the sinusoidal vibration of displacement amplitude 10 mm and frequency 5 Hz (Example 2.1) is shown on the nomograph as point G, and approximate values of the other two quantities, which were calculated to be velocity 314 mm/s and acceleration 9860 mm/s^2, can be read directly off the appropriate scales of the nomograph.

The slope of the lines of constant displacement can be checked by taking, for example, the line for 10 mm displacement ($Z = 10$ mm). The points B and C where this line cuts the vertical lines for frequencies of 1 and 10 Hz (i.e., $\omega = 2\pi f = 6.28$ and 62.8 rad/s) correspond to velocity $\omega Z = 62.8$ and 628 mm/s.

The slope of the constant acceleration lines can be checked by calculating the velocity at points D and E, for example, where the acceleration line for 10^4 mm/s^2 cuts the lines for frequency $f = 1$ and 10 Hz. The velocity at these points is $\omega Z = \omega^2 Z/\omega = 10^4/6.28$ and $10^4/62.8 \approx 1590$ and 159 mm/s.

The major lines of the diagonal grid intersect on the major horizontal lines. For example, at the point F, the displacement is $Z = 10$ and $\omega^2 Z = 1000$.

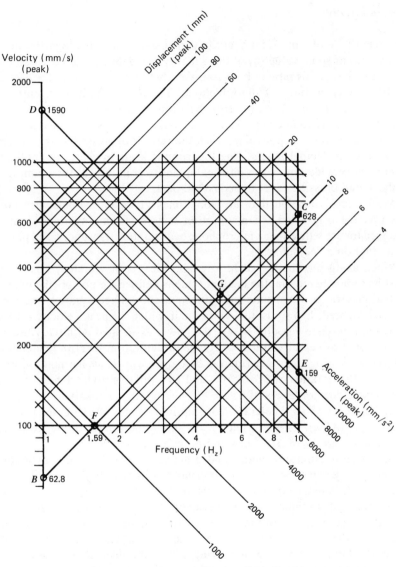

Figure 2.3. Nomograph for sinusoidal vibration.

Therefore $\omega = 10$ rad/s and the velocity $\omega Z = 100$ mm/s. The frequency at the point F is $f = \omega/2\pi = 10/2\pi = 1.59$ Hz.

The axes of the nomograph in Fig. 2.3 are annotated for peak values of the physical quantities concerned. Nomographs are commonly used to present data in terms of the root mean square (rms) value, which is defined later in this chapter. For example, the nomograph used later (Fig. 4.4) presents vibration criteria in terms of the rms values.

Decibel Scale

The magnitude of a physical quantity can be expressed on a logarithmic scale as the ratio of some value (q) of the quantity to a defined reference value (q_0). The logarithm of this ratio to base 10 is the "bel" which is named for Alexander Graham Bell, inventor of the telephone. The *deci*bel unit, one tenth of the bel, is $10 \log_{10} (q/q_0)$. Similarly, the ratio of two values (q_1 and q_2) of a physical quantity can be expressed in decibels: if each is expressed in decibels with the same reference level q_0, then the ratio is expressed as the difference between their decibel levels. The logarithmic scale in useful in applications where a very wide range of a physical quantity is involved; for example, that of the sound pressure level in acoustical data.

Mechanical vibration data are sometimes given in decibels. Presumably this practice originated when vibration meters became available that were designed for use with either a microphone or a vibration transducer, and presented the output on a decibel scale.

When using the decibel scale to express levels of vibratory quantities, it must be remembered that the decibel scale is defined for power-like quantities such as electric power, sound-pressure squared, voltage squared, and particle velocity squared. In a sinusoidal vibration in which the peak velocity is denoted by V and reference velocity by V_0, the power associated with the vibration is proportional to velocity squared. Hence the vibration level in decibels is $10 \log_{10} (V/V_0)^2$, which may be written as $20 \log_{10} (V/V_0)$. Thus, harmonic vibrations having velocity ratios of 2, 10, 10^2, and 10^3, relative to the defined reference value, have decibel levels of 6, 20, 40, and 60 dB, respectively.

In the same way an acceleration of amplitude A may be expressed in decibels as $20 \log_{10} (A/A_0)$. The reference level should be stated; reference levels in common use are $V_0 = 10^{-9}$ m/s and $A_0 = 10^{-6}$ m/s^2.

In vibration-isolation theory and practice, the performance of a seismic mounting is described in terms of a transmissibility ratio (e.g., ratio of vibration amplitude of isolated equipment to that of site) and therefore can be expressed in decibels but, in the author's experience, there is no good reason for doing so. The range of the ratio that is of practical importance is not wide enough to require the use of a logarithmic scale and, when used as a unit of the vibration amplitude at one measuring point, the decibel introduces unnecessary complication: for example, why say that a vibration velocity is "120 dB (rel. $V_0 = 10^{-9}$ m/s)" instead of simply "1 mm/s"? The use of the decibel scale is unattractive also if mechanical vibration data are to be presented on a nomograph, for it is obviously preferable to mark the axes with immediately recognizable units of displacement, velocity, and acceleration (e.g., mm, mm/s, mm/s^2) than to mark all three axes with decibel units identifiable only through the reference level as in the example just given for velocity.

Periodic Vibration

A vibration is periodic if its waveform is repetitive. In the example in Fig. 2.4, the value z_1 of the displacement of point P from the datum position O at any instant t_1 recurs at intervals of T, $2T$, . . . seconds, where T is the time taken for one complete cycle of the vibration. Periodic vibration is generated by engines, compressors, pumps, and other kinds of continuously running machinery. Sinusoidal vibration is a particular form of periodic vibration.

The pattern of motion of a point making a periodic vibration may be unsymmetrical about the datum, which is commonly taken to be the position that the point occupies when the system is at rest. If the maximum displacement on one side of the datum is different from that on the other, we use the terms positive or negative peak displacement.

A point oscillating with periodically varying displacement obviously must experience velocity and acceleration that also are periodic. The waveforms of velocity and acceleration can be derived graphically or mathematically from the displacement waveform. In practice the displacement, velocity or acceleration of a vibrating body may be measured by using a transducer that generates the required electrical analogue (see section on Vibration Measurement). Or, by using an acceleration transducer in conjunction with electrical integration circuitry, the velocity or displacement can be measured at the throw of a switch.

As the waveform of a periodic vibration can take a great variety of shapes, the *peak* value alone is not a very informative way of describing the vibration. The *average* value is also not informative because for sinusoidal vibration it is always zero regardless of the magnitude of the excursion to the positive and negative side of zero, and for most other periodic vibrations the average value differs little from zero. A more useful measure is the *absolute average* value,

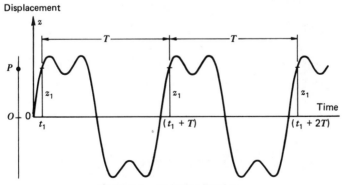

Figure 2.4. Periodic vibration.

which is the average value obtained by ignoring the sign of the negative values; for sinusoidal vibration the absolute average is 0.637 of the peak value.

If each value of the displacement z in Fig. 2.4 is squared, a curve of z^2 versus time results. The squaring yields all positive values, and the square root of the average value of z^2 is the *root mean square (rms)* value. This is the most useful and generally used measure of the magnitude of vibration because, in addition to taking the waveform into account, it is directly related to the energy associated with the vibration and is meaningful also for nonperiodic or random vibration. For sinusoidal vibration the rms value is $1/\sqrt{2}$ or 0.707 of the peak value.

The most remarkable and useful feature of a periodic vibration is that it can be regarded as being composed of sinusoidal or harmonic components comprising the fundamental or lowest frequency component and a number of others having frequencies that are integral multiples of the fundamental frequency. Any periodic vibration can be analyzed, either mathematically or by the use of electronic instruments, to determine these components, which are named *Fourier*

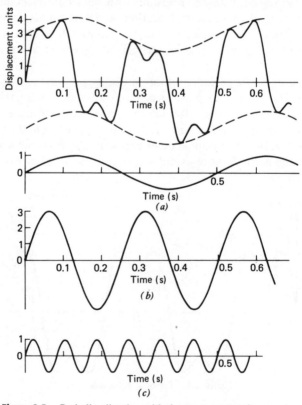

Figure 2.5. Periodic vibration with three component frequencies.

components for Jean-Baptiste Fourier who explained the underlying mathematical theory more than a century ago.

Sometimes it is easy to identify Fourier components by looking at the waveform. For example, the waveform shown at the top in Fig. 2.5 has the following three components, which are shown below it in the figure.

(*a*)   The presence of this lowest-frequency component is disclosed by the fact that the waveform lies within an envelope, shown by the broken lines, which completes one cycle in 0.5 s and hence has a frequency of 2 Hz. This is the fundamental or first harmonic.

(*b*)   The major and most obvious component makes two cycles in 0.5s, and hence is the second harmonic, having a frequency of 4 Hz.

(*c*)   The highest-frequency component is evident as the undulation or ripple which completes 6 cycles in 0.5 s; this is the sixth harmonic, having a frequency of 12 Hz.

A periodic vibration can be described by its line spectrum. Figure 2.6*a* shows the line spectrum of the vibration in Fig. 2.5. Each component is shown by a line of length representing its amplitude, located on the frequency scale at the frequency of that component.

The Fourier components of vibrations met in practice are not always obvious in the waveform. Even when there are only a few components, the shape of the waveform can change markedly with change in the phase or relative positions of the components along the time axis. In practice, frequency is usually of more interest than phase. The important components are found by electrical filtering methods, which produce a record on which sharp peaks identify the components, the abscissa position giving the frequency and the height giving the amplitude. The record for the vibration shown in Fig. 2.5 would look like that sketched in Fig. 2.6*b*. Frequency analysis is discussed later under the heading Vibration Measurement.

Periodic vibrations are associated with many phenomena in mechanical and electrical engineering and other fields, as well as in vibration and acoustics. Explanations of Fourier analysis, expressed in varying degrees of complexity, can be found in text and reference books in many fields and in mathematical texts. For the purpose of the present book, it is sufficient to note that the components can be found for all periodic vibrations, even those having square and sawtooth waveforms (e.g., see Bishop, 1979, pp. 14–19).

Knowledge of the Fourier components is essential in vibration investigations. The component frequencies are clues to identification of the vibration sources, and also they provide a basis for predicting the effects that the vibration may have on equipment and structures.

A special kind of rectilinear vibration occurs when the vibration is the result of the combination of two sinusoidal components that differ in frequency by

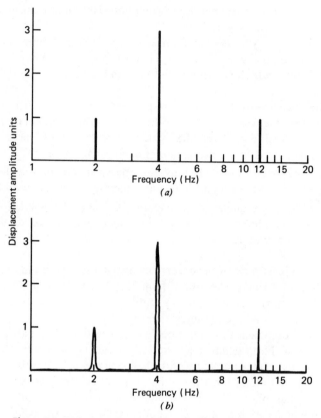

Figure 2.6. Line spectrum of the periodic vibration in Fig. 2.5.

an amount that is small in comparison with the frequency of either component. The waveform of the resulting vibration looks like a sinusoid whose amplitude is cyclically increasing and decreasing (*beating*). The *beat frequency* is equal to the difference between the two component frequencies. Beats occur, for example, when two machines having the same nominal rotational speed are running with slightly different speeds. The effect may be audible; for example, passengers in a twin-engined aircraft may hear beating for a short period after take-off until the pilot synchronizes the engines to have exactly the same rotational speed, in which condition the beat frequency is zero.

Random Vibration

Suppose that during a vibration test in which all conditions under the control of the operator are kept constant, successive waveform samples of the vibration at a particular point on the test object are recorded. If the vibration is periodic, each sample will look like the others and any one sample can be taken as truly representative of the vibration. If the samples look different from one another, the vibration being recorded is "random" and no one sample (e.g., Fig. 2.7a) is adequate to represent the vibration. Therefore, we must resort to statistical methods to describe the vibration in terms of some kind of average of the characteristics of a sufficiently large number of samples of the vibration. We need some statistical measure of the magnitude of the vibration and some way of describing its frequency content.

For the magnitude, methods have been developed whereby, for a given random vibration, a curve can be derived to show the probability that the instantaneous value of the magnitude will be within a certain small part of its range. However, in many engineering problems it is adequate and preferable to characterize a random vibration simply by the rms value.

For the frequency content, whereas a periodic vibration has Fourier components that can be presented as a line spectrum (Fig. 2.6), the energy in a random vibration is spread over all frequencies; that is, it has a continuous spectrum as shown in Fig. 2.7b. One can think of the spectrum of a random vibration as being composed of a large number of narrow bands, such as that shown crosshatched, packed closely together as in the figure.

Unlike a periodic vibration, which is conveniently described in terms of the rms value of its harmonic components, a random vibration is best described in terms of its mean square value. Thus, the height of each of the narrow bands

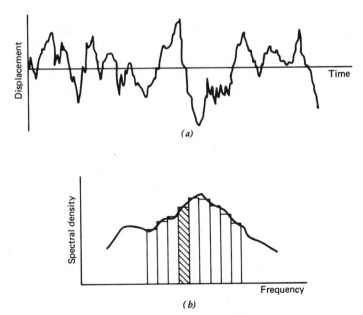

(a)

(b)

Figure 2.7. Random vibration.

in Fig. 2.7*b* represents the mean square value of the magnitude for that band. A presentation of this kind is called the *mean square spectral density* or *power spectral density*.

In practice, the process of deriving such a spectrum is equivalent to passing the vibration signal, that is, the electrical output of a vibration transducer, through a large number of filters, each of which passes only a narrow band of frequencies centered on a particular frequency, and which together cover the frequency range of interest. The mean square value of the output of each filter is plotted as part of the power spectral density curve.

Other, more sophisticated statistical techniques have been devised for the investigation of problems involving random vibration. *Autocorrelation* reveals periodicity concealed in mixed random/periodic vibration. *Cross correlation* discloses whether signals from two separate vibrating objects, whose waveforms may look different, may have some frequency correlation. For example, one body may be shown to be receiving vibration from the other, or both may be shown to be responding to another source. An example of the use of a correlation technique to identify a vibration source is given later, in the section on Vibration Measurement (Example 2.4).

Transient Vibration

A transient vibration occurs when the shape or configuration of a mechanical system is suddenly changed as, for example, when it is struck, bumped, or otherwise subjected to an impulse, or when it is released after being displaced from its rest position. The system vibrates for a short time then comes to rest again. Familiar examples are the vibration of a floor when something falls on it, and the vibration of a plucked string. The amplitude decays because energy is dissipated by internal friction in the vibrating materials, or by external friction, or some other mechanism. This dissipation of energy is referred to as *damping*.

A transient is usually described by its waveform of displacement, velocity, or acceleration versus time. The magnitude can be given as the peak, average, or rms value, as with other vibrations. The predominant frequency is usually obvious, and other components may be evident on inspection of the waveform, as with periodic vibration. If not, it is necessary to assume that the transient has a continuous spectrum and to use the power spectral density to describe the transient.

If the waveform record shows a predominant frequency and continuously decreasing amplitude, it can be used, in the manner detailed below, to estimate the natural frequency and damping of the body or element whose transient response is recorded. It will be seen in later chapters that the natural frequency and damping of structures, equipment, and seismic mountings are of great importance in vibration isolation.

Damped Harmonic

The most familiar transient is the damped harmonic, an example of which is shown in Fig. 2.8. Such a transient represents the free vibration of the ideal mass–spring–damper system, which is discussed later (Chapter 6). In effect, the waveform is a sinusoid whose amplitude is decreasing within the exponentially decaying envelope shown by the broken lines. Transients of the purity illustrated are more common in technical literature than in practice. Nevertheless, the damped harmonic is very useful as an idealization of the transients that occur in practical situations.

The natural frequency and damping of a structural or other element can be evaluated from a record of the free vibration of the element in the following way. A record of the transient is made by using a suitable measurement technique (see section on Vibration Measurement), and the transient is assumed to be of damped harmonic form, as in Fig. 2.8 (see also Example 2.4 and Fig. 2.16). The natural frequency is evaluated by counting the number of cycles completed in a given time.

The damping is evaluated from the rate of decay of the amplitude (e.g., see Den Hartog, 1956, pp. 37–40) and is expressed in terms of the *damping ratio*, which is defined as the ratio of the actual to the *critical damping*. The latter is defined as the least amount of damping that would suppress the free vibration so that the element, if released from a displaced position, would return without "overshoot" to its rest position.

Example 2.2.   Determine the damping ratio of the damped harmonic in Fig. 2.8.

The rate of decay of the transient is found by measuring the heights (Z_0, Z_1, . . .) of successive peaks on the record. The damping ratio (ζ) is then evaluated as

$$\zeta = -\frac{1}{2\pi n}\ln\left(\frac{Z_n}{Z_0}\right) \tag{2.3}$$

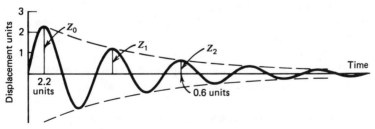

Figure 2.8.   Damped harmonic vibration.

where Z_n is the height of the peak after n cycles of the vibration. This expression gives an approximation adequate for design purposes relating to systems having damping ratio less than about 0.2.

Referring now to Fig. 2.8, we see that in the first two cycles ($n = 2$) the peak amplitude decays from $Z_0 = 2.2$ to $Z_2 = 0.6$ units. Substituting in Eq. (2.3) gives

$$\zeta = -\frac{1}{4\pi} \ln\left(\frac{0.6}{2.2}\right)$$

$$\approx 0.1$$

The natural frequency derived from a record of a damped harmonic is the *damped natural frequency*, which differs from the natural frequency that the system would have in the absence of damping. The difference depends on the amount of damping and is unimportant in the applications dealt with in this book ($\zeta \leqslant 0.2$).

Rotational Vibration

Every point on a body that is vibrating in pure translation makes a rectilinear vibration, which may take any of the forms discussed. Before discussing more general vibratory motion, which involves both translation and rotation, we define pure rotational vibration.

Consider a body (Fig. 2.9) oscillating about the axis Ox, perpendicular to the page. The instantaneous position of the body, shown as a solid line, is defined in terms of the angular displacement (α) of the body from its rest or datum position, which is shown as a dashed line. All the foregoing discussion of the basic kinds of rectilinear vibration is applicable to angular vibration if the cyclically varying quantity is taken to be the angular instead of the rectilinear displacement.

Rotational vibration may be associated with torsional strain, in which case the angular displacement in a particular plane represents that of one part of a body (e.g., a shaft) relative to another part of that body. Torsional vibration is an important consideration in the design of rotors and shafts transmitting power, and in the design of aircraft, bridges, and other structures that must withstand torsional strain associated with aerodynamic and other excitations.

In this book the rotational vibration of specific interest is that of a seismically mounted machine or equipment. A vibrating machine may generate unbalanced forces and couples, which tend to cause rotational as well as translational motion. We now define the terminology used to describe the three translational and the three rotational components of motion of a rigid body.

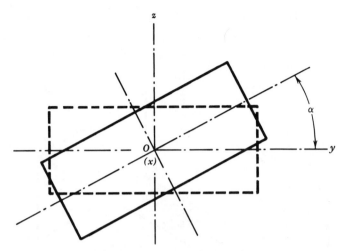

Figure 2.9. Rotational vibration.

Rigid-Body Vibration

A rigid body occurs only in the mind as a theoretical concept. Any material body must alter its shape to some extent when it is vibrating. Nevertheless, it is convenient and permissible to regard a body as rigid when the elastic vibrations of the body are either negligible in comparison with its rigid-body vibrations, or unimportant in the context of the particular problem.

We refer to Fig. 2.10 and assume the body to be rigid. If the vibration is purely rectilinear the motion of any particle of the body will be identical with that of any other particle. Therefore, the motion can be described in terms of the rectilinear vibration of the center of gravity (c.g.), as though the whole mass of the body were concentrated at the c.g. In general, a rectilinear vibration will not be purely vertical or horizontal, so it is described in terms of the components of the motion of the c.g. referred to the axes Ox, Oy, Oz, having their origin at the position occupied by the c.g. when the body is in the rest or datum position.

In dealing with the angular vibration of the whole body regarded as a rigid mass we make use of the fact, which is shown in textbooks on mechanics, that through any point in a rigid body there is a set of rectangular axes, called the *principal axes*, such that the moment of inertia is a maximum about one axis, a minimum about another, while that about the remaining axis is, of necessity, intermediate between those about the other two. In theoretical analyses of the rotational oscillations of a rigid body it is advantageous to use the c.g. of the body as the origin of the set of rectangular axes of reference and to adopt the principal axes through the c.g. as the axes of reference.

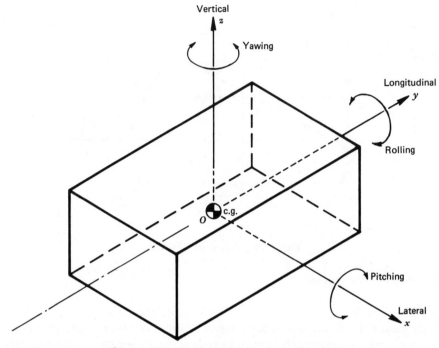

Figure 2.10.   Modes of vibration of a rigid body.

Angular oscillation is described in terms of the motion about the principal axes. The terminology of ship motion is commonly used: referring to Fig. 2.10 pitching is the oscillation about Ox, rolling is that about Oy, and yawing about Oz. When the vibration involves angular vibration about all three axes at the same time, the motion is described in terms of the components of angular motion referred to the three axes. If the body has translational and rotational vibration at the same time, the motion is described in terms of the translational components of the c.g. and the rotational components about the principal axes.

In the theoretical treatment of the design of vibration-isolating mountings, consideration of combined rotational and translational motion in all its generality is quite complicated. For the practical purposes dealt with in this book we show later that the treatment need be no more complicated than that involving combined rotation and translation in one plane; for example, motion parallel to the vertical plane yz involving translation of the c.g. parallel to Oy combined with rotation (pitching) about Ox.

Elastic Vibration

The elastic vibrations of a body are of the three basic kinds indicated in Fig. 2.11: (a) longitudinal tension and compression, (b) flexural or bending,

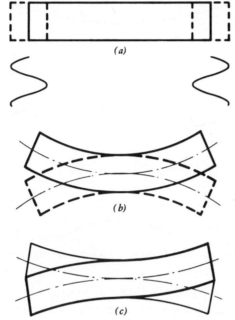

Figure 2.11.   Longitudinal, flexural, and torsional modes.

and (c) torsional or twisting. A body vibrating in any one of these ways can do so in different modes. For example, Fig. 2.12 shows the first three modes of flexural vibration of a simply supported beam. Analogous modes can occur in longitudinal and torsional vibration.

In this book flexural vibration is important because of its relevance to the design of installations on suspended floors. The modal shapes and natural frequencies of an existing floor can be found experimentally (e.g., see Example 2.3 and Fig. 2.14). Calculation of the natural frequencies and modal shapes of a floor from the structural drawings may be very complicated if the floor is unsymmetrical in plan shape and in the arrangement of the supporting beams, columns, and walls.

Fortunately, we need not venture into this more difficult field because, for the purposes of this book, we can replace a system having distributed properties with an equivalent "lumped mass" system. For example, in relation to the vibration isolation of machinery on a suspended floor (Chapter 8) the floor is represented by an equivalent mass–spring system as discussed in Chapter 6. Of course, in doing this we must be sure that those properties of the actual floor that are important in the context of the problem are adequately represented by those of the substituted equivalent system.

There is no reason why a body that is part of a vibrating system should not be considered to be rigid for some aspects of the design and nonrigid for others.

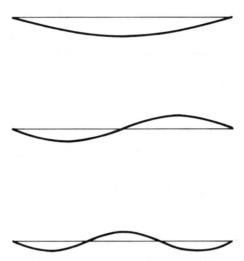

Figure 2.12.   Fundamental, second, and third flexural modes of a simply supported beam.

For example, one may treat a building as a rigid body for the purpose of describing its rocking vibration in response to wind gusting or an earthquake, and as a body having distributed elasticity in describing the responses of particular parts of the building.

Another example is met in the design of a mounting for vibration-sensitive equipment. The equipment is installed on an "inertia block" which rests on resilient isolators (Chapter 5). The natural vibrations of the entire mass on the isolators are described as those of a rigid mass, but for the purpose of designing to ensure adequate stiffness of the block to preserve the alignment of the equipment, the vibration of the block must be described as that of an elastic beam capable of vibrating in flexure and torsion (e.g., see Example 5.9).

VIBRATION MEASUREMENT

Vibration measurement is helpful in many ways in relation to vibration control. For example, measurement of the vibration generated by a machine can show if the vibration is severe enough to damage the machine itself or the supporting structure, or disturb sensitive equipment or humans in the vicinity. If necessary the machine may be dynamically balanced *in situ* by techniques involving vibration measurement. A vibration survey can identify the least-disturbed site for sensitive equipment.

If a decision is made to provide a seismic mounting, vibration measurements can give the necessary information about the vibration excitation, the dynamic characteristics of the supporting structure, and those of the equipment to be isolated. Vibration measurements may be used also to investigate the performance of the completed seismic mounting.

For these purposes, the range of frequency to be measured may be from a fraction of 1 Hz, the swaying frequency of a tall building, to some hundreds of hertz, which covers the important frequencies associated with the operation of machinery and the natural frequencies of the responsive parts of sensitive equipment. The displacement range of interest in normal practice is from less than 1 μm, in relation to installations of very sensitive equipment, to some hundreds of millimeters for the motions of large structures and tall buildings (see Fig. 1.1).

The basic principles of vibration measurement are indicated in the following brief review of the development of vibration-measuring instruments, from the cumbersome devices used some 50 years ago to the small transducers and versatile electronic instrumentation used today. The scope of the discussion is limited to aspects of measurement that are relevant to normal practice in vibration isolation: no reference is made to advanced instrumentation and techniques used in research and measurement laboratories.

The theory of vibration measurement is well documented (e.g., see Den Hartog, 1956, pp. 57–69; Harris and Crede, 1976, Chaps. 12–18). Reviews of the technical literature on shock and vibration measurement appear regularly in the *Shock and Vibration Digest*; the most recent at the time of writing is by Plunkett (1982). A useful practical guide on how to manage and conduct vibration measurements and analyses of large structures is given by Dorien-Brown and Meldrum (1976). Some excellent literature on vibration measurement is available from the major manufacturers of vibration instrumentation.

Development of Vibration Transducers

The motion of a point on a vibrating object is measured in terms of its cyclically varying position relative to some reference or datum position. For example, if a structure supporting an unbalanced machine is obviously vibrating through a distance of some millimeters relative to an adjacent nonvibrating building, the displacement can be read directly on a steel rule or other suitable scale held on the fixed or datum building. In practice this situation occurs very rarely: usually no fixed point of reference is available, and the displacement is too small to be measured without magnification of some kind.

An artificially "fixed" point can be provided by using a seismic suspension. In seismographs for recording earthquakes, this consists of a large mass on a soft spring suspension supported on a base or frame fixed in the ground. The mass remains almost stationary when the frame vibrates at frequencies appreciably higher than the natural frequency of the suspension. The relative displacement between mass and frame is taken as a measure of the absolute vibration of the frame. The seismograph is the classic example of a vibration isolating system: the soft suspension isolates the seismic mass from the ground vibration, and hence the mass can be regarded as a stationary datum for the ground vibration measurement. This behavior of a spring-supported mass is discussed in detail later, in the context of vibration isolation (e.g., see discussion in relation to Fig. 6.2).

As needs arose for portable instruments to measure vibration in engineering activities, smaller seismic suspensions were devised in a fascinating variety of arrangements and configurations, using a helical spring, flexure strips, or a pendulum to support the seismic mass. Whereas the seismic mass in the earlier portable instruments was supported on soft springs, another type was developed using stiff springs. The significance of this is discussed later in the section on Accelerometers.

In all these instruments a magnified indication of the vibration was derived from a device connected between the seismic mass and the frame. These devices have taken innumerable forms, including mechanical linkages; tilting mirrors and optical levers; and electromagnetic, inductive, capacitive, and other kinds of electrical transducer used in conjunction with electrical amplifiers, bridges, and the like. The resulting indication was displayed as a pointer-on-scale or optical image reading, or as a waveform on paper, film, or oscilloscope.

Most of the earlier instruments described in the literature are now of only historical interest. Generally, the calibration factor was calculated, or based on a static magnification test, in the expectation that the instrument, however complicated and massive its components and linkages, would behave as a single-degree-of-freedom system (see Chapter 6) throughout (usually unstated) ranges of amplitude and frequency. It is suspected from recent calibrations of some of the vintage instruments that the records they produced may have been more informative of their own shortcomings than of the characteristics of the vibration they were supposed to be measuring.

Vibration Measurement Today

Today most vibration measurements are made with small seismic transducers (pickups), used in conjunction with amplifiers and other signal conditioning equipment, and some means of displaying the output signal. The main attractions of a small transducer are that it is easier to connect to the vibrating object, it is less likely to modify the motion being measured, and it is easier to calibrate than the rather large instruments formerly used.

The most important characteristics of a transducer are the sensitivity, the frequency response (particularly the "flat" range in which the indicated amplitude is constant for constant applied amplitude), and the linearity of the relationship between indicated and applied amplitude at a given frequency. Other considerations are sensitivity to vibration in directions other than that intended to be measured, and sensitivity to the environment (temperature, electrical fields, etc.). The associated electrical amplifiers, filters, integrators, and recorders should have adequate frequency response, gain, and linearity to ensure that they cause no distortion of the output signal.

The entire system from transducer to output should be calibrated for sensitivity, frequency response and amplitude response, and for any other of the abovementioned characteristics relevant to the conditions of intended use.

In modern transducers the seismic mass is supported on flexure strips, diaphragms, or other form of spring element designed to deflect in only one direction, that of the "sensitive axis" of the transducer. The relative motion between the seismic mass and the body of the transducer is detected and magnified electrically to produce an output signal, which is a measure of the axial vibration of the body of the transducer. The seismic suspension is designed to have frequency and damping properties appropriate to the intended application of the transducer.

Displacement Transducers

If the natural frequency is below the range of frequency to be measured (i.e., soft suspension), the seismic mass behaves like that of the seismograph mentioned earlier, and the relative displacement of the mass is about the same as the absolute displacement of the body of the transducer. Seismic transducers with output proportional to displacement are seldom used today. Most measurements are made with velocity or acceleration transducers, from which displacement, when required, is obtained by electrical integration.

When it is necessary to measure or monitor relative displacement, for example, the radial clearance between a rotating shaft and its bearing, proximity (nonseismic) transducers are used. In relation to the design of seismic mountings for sensitive equipment, proximity transducers are used to measure the relative vibration between a critical or responsive part of the equipment and its base or frame, for purposes that are discussed in Chapter 9 (see also Examples 5.12).

Velocity Transducers

A transducer with a soft suspension may be designed to function as a velocity transducer by connecting, between the seismic mass and the body of the transducer, an electromagnetic element that generates an output signal proportional to the relative velocity. Since velocity is the parameter now generally correlated with structural damage (Chapter 4) ground and structural vibration measurements are made with velocity transducers, or with accelerometers used in conjunction with electrical integration to give velocity data. For these applications in which the size of the transducer is unimportant, relatively large yet portable seismometers, which are very sensitive velocity transducers, are used.

For ground vibration measurement, care must be taken to ensure that the vibration transducer is satisfactorily coupled to the soil; otherwise, the output signal may involve serious error resulting from the response of the transducer relative to the ground (e.g., see Example 2.3 and Nolle, 1978).

Velocity transducers are now commercially available in small sizes, though not as small as accelerometers because the suspension required to achieve the low natural frequency necessary for a velocity transducer takes more space than the stiff suspension used in an accelerometer.

Accelerometers

The transducer can be designed to function as an acceleration transducer or "accelerometer" by supporting the seismic mass on a stiff suspension such that the natural frequency is above the range of vibration frequencies to be measured. The most common accelerometer consists of a mass fixed on a piezoelectric material fixed on a base that is attached to the vibrating surface. The oscillatory loading of the piezoelectric element, which functions as both the stiff spring and the source of the electrical signal, produces an alternating electrical charge and a corresponding voltage, which are related to the imposed acceleration. Accelerometers are extensively used for measuring mechanical shock, for which purpose extremely small transducers (e.g., a few grams mass) having very high natural frequencies have been developed.

An important variation of the seismic transducer is the servo accelerometer, in which a soft suspension is used but the output signal is not derived from the relative motion between the seismic mass and the body of the transducer. Instead, an electrical feedback current maintains the seismic mass in a null position, and the output signal, which derives from this current, is proportional to the acceleration. The seismic suspension is mechanically soft and is, in effect, stiffened by the electrical feedback system. Servo accelerometers are particularly useful for measuring vibration of very low frequency (e.g., a fraction of one hertz). Servo accelerometers are commercially available with frequency response "flat" down to zero frequency.

Multichannel Measurements

In many practical applications vibration must be observed at two or more points simultaneously. For example, if the effectiveness of a vibration isolating mounting is being measured, one transducer may be placed on the isolated equipment and another on the adjacent floor, and their outputs simultaneously recorded under given conditions of excitation. Additional transducers are used if observations are to be made at several points and in various directions on both the equipment and the floor. The outputs from all the measuring channels may be recorded on a direct writing, multichannel oscillograph, or on tape for later examination on playback.

Multichannel measurements are used also to determine the modal shapes of vibrating structures. Transducers are arranged on the structure in positions and orientations chosen to suit the modes to be observed, and the outputs recorded on a multichannel direct-writing oscillograph. By examining the waveforms and noting which transducers are vibrating in phase and which antiphase, the mode of vibration of the structure, whether flexural or torsional, fundamental or higher mode, can be determined. The method is discussed and illustrated in some detail in a description of a vibration study of a large radiotelescope (Macinante et al., 1967). The following more recent example of the use of multichannel instrumentation and techniques refers to an investigation by Goldberg and Drew (1980).

Example 2.3.   Measurements were made to determine the response of a modern high-rise building to ground vibration caused by blasting. The building, a 10-story reinforced concrete structure, is situated 1.4 km from the site of blasting that was in progress to deepen the local harbor.

The ground vibration in the vicinity of the building was measured by using three independent velocity seismometers: two horizontal, the other vertical. The peak particle velocity (ppv) of ground vibration at any instant is the resultant of the instantaneous values of the three components. Each horizontal seismometer was supported on a peg driven into the ground. The peg, which

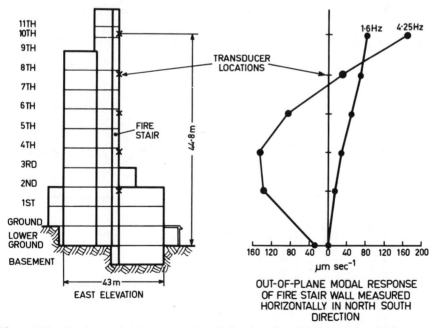

Figure 2.13.   Fundamental and second modes of vibration of a tall building (after Goldberg and Drew, 1980).

is of cruciform section, with a flat top for the seismometer, and tapered toward the bottom, was designed to permit effective coupling of the seismometer to the ground. It was confirmed experimentally that the coupling was satisfactory over an adequate frequency range (up to 50 Hz). The vertical seismometer was buried in the ground as a package having about the same density as the soil.

The flexural response of the building as a whole, in the N–S and E–W planes, was determined with a group of accelerometers with their sensitive axes horizontal and in a common vertical plane at different levels as shown in Fig. 2.13, which shows the observed fundamental flexural mode (1.6 Hz), and the second mode (4.25 Hz), in the N–S plane.

The multichannel technique was used also to observe the mode of vibration of a typical floor and wall. Figure 2.14 shows part of the mode shape of the response of a floor at a frequency of 6.75 Hz.

In the light of the results of the investigation, Goldberg and Drew point out that codes of practice which use the ground vibration velocity near the building as a criterion of acceptable vibration for human comfort and building damage (see Chapter 4) may need modification to take into account the magnification of vibration resulting from dynamic response of the building as a whole and of its floors and walls.

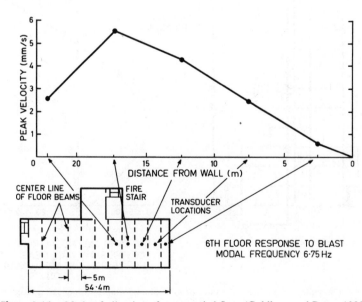

Figure 2.14. Mode of vibration of a suspended floor (Goldberg and Drew, 1980).

Frequency Analysis

When diagnosis of objectionable vibration is necessary as a basis for deciding the remedial action to be taken, frequency analysis is essential. As illustrated earlier in this chapter, any periodic motion can be regarded as a combination of a number of sinusoidal components with harmonically related frequencies: the more complicated the waveform, the more components necessary to describe it. If the component frequencies are discovered by making a vibration analysis, it is usually a simple matter to relate them to particular shaft rotational speeds, gear meshing frequencies, electrical excitation frequencies, and so on, and hence to identify particular frequencies with, say, shaft unbalance, misalignment, or gear noise.

A few decades ago, analysis had to be done by laborious graphic reduction of samples of the recorded waveform. An experienced observer could often evade this task by making a visual inspection of the recorded trace and its envelope and thereby extracting enough information to identify the fundamental and two or three more obvious components (e.g., see Manley, 1945, and the example given in Fig. 2.5). There was some interest at the time in mechanical analyzers, but the significant developments have come through the application of electrical filtering methods.

Today all frequency analysis is done electrically. When the vibration frequencies of interest are within a well-defined band, individual low-pass, band-pass or high-pass filters can be used which reject frequencies outside the range of interest. A more versatile analyzer is the sweep-frequency type in which a very narrow band-pass filter is automatically swept through a frequency range and the results presented as a spectrum showing the relative amplitudes of the various frequency components. Sweep frequency analysis is adequate if the vibration remains unchanged during the few minutes it takes to sweep through the frequency range. The most important development is the real-time analyzer, which displays the frequency spectrum instantaneously and continuously. This permits analyses to be made much more rapidly, and has the advantage that it reveals changes in frequency content, due to variations of the vibration source, which could have been missed in the course of a sweep analysis.

The following summary of an investigation of the influence of structural vibration on the quality of the optical image in a solar magnetograph (Goldberg and Dorien-Brown, 1972) provides an example of the use of frequency analysis and statistical correlation techniques.

Example 2.4.   The CSIRO Solar Observatory at Culgoora, N.S.W., Australia, operates a magnetograph, which is situated at the top of a tower 15.2 m high, and of rectangular section 4.9 × 6.7 m. The structure is of reinforced concrete with all floors, columns, beams, footings, and stairs formed as integral parts of the structure. With any optical telescope, perturbation of the image, resulting in degradation of quality of photographs, is usually attributable to atmospheric inhomogeneity or "seeing." However, structural vibration also may contribute to disturbance of the image. The object of the investigation was to determine the role of vibration.

A signal related to the image movement was derived from a transducer devised from a curved slit, with its center line coincident with the edge of the solar image, and a photodetector. The angular vibration of the optical axis was monitored with two accelerometers attached at each end of the spar of the telescope. The signals from the four accelerometers and the photodetector were recorded on a multichannel direct-writing oscillograph, together with a timing trace from a crystal oscillator. A typical record is shown in Fig. 2.15. To the right of each waveform sample is the plot of the power spectral density.

The signal from the photodetector shows peaks at 1.0 and 4.5 Hz. The signals from accelerometers 2 and 3, which are at opposite ends of the spar,

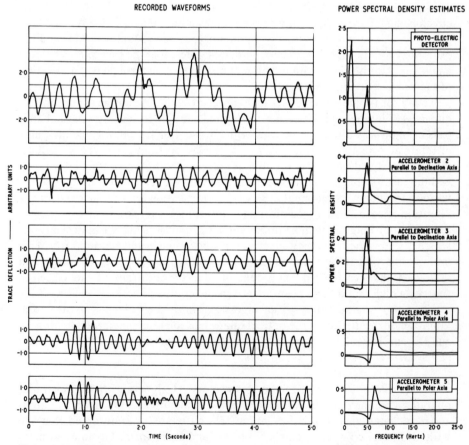

Figure 2.15. Waveforms and power spectral density estimates of multichannel vibration record (Goldberg and Dorien-Brown, 1972).

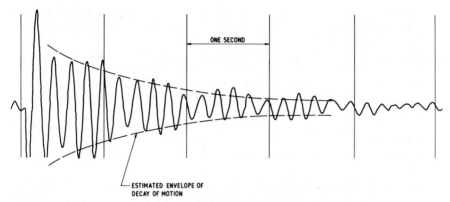

ONE SECOND

ESTIMATED ENVELOPE OF
DECAY OF MOTION

Figure 2.16. Record of free vibration of a building (Goldberg and Dorien-Brown, 1972).

also show a signal peak at 4.5 Hz, and accelerometers 4 and 5 show a peak at 6.5 Hz. The 1.0-Hz component was attributed to an instability of the servo-mechanism, induced by the particular load on the spar during the observations, and was not considered further. Further analysis using "cross correlation" showed highly significant correlation between image motion and angular oscillation of the spar in the plane containing the sensitive axes of accelerometers 2 and 3, and reduced correlation with 4 and 5. With the telescope tracking, the angular vibration of the spar had a predominant frequency of 4.5 Hz about the polar axis, and 6.5 Hz about the declination axis.

The mode of vibration of the building was investigated with an accelerometer on the telescope, and seismometers oriented in the N–S direction at successive positions on a vertical line. The results show that the building was vibrating as a vertical cantilever on a compliant foundation. Similar behavior was observed in the E–W direction. The frequency of the major component in both the N–S and E–W directions was in the range 5–6 Hz. From records of the free vibration of the tower in response to transient excitation, the natural frequency was 5.9 Hz N–S and 5.2 Hz E–W, and the damping ratio was in the range 0.02–0.04. A typical record is given in Fig. 2.16.

This vibration analysis showed that the frequencies of vibration of the building are close to the natural frequencies of vibration of the spar, and that the resulting angular vibration of the spar contributes to the disturbance of the solar image.

REFERENCES

Bishop, R. E. D. (1979), *Vibration*, 2nd ed., Cambridge University Press, Cambridge, England.

Den Hartog, J. P. (1956). *Mechanical Vibrations*, 4th ed., McGraw-Hill, New York.

Dorien-Brown, B., and B. H. Meldrum (1976). Measurement of the Vibration of Large Structures, *Vibration and Noise Control Engineering Conference*, Institution of Engineers, Australia, National Conference Publication No. 76/9, pp. 6–10.

Goldberg, J. L., and B. Dorien-Brown (1972). Image Motion in the Culgoora Solar Magnetograph—the Role of Vibration, *Publ. Astron. Soc. Pac.*, **84**(500) 534–540.

Goldberg, J. L., and P. Drew (1980). The Response of High-Rise Buildings to Ground Vibration from Blasting—An Experimental Investigation, *10th International Congress on Acoustics*, Sydney, Australia.

Harris, Cyril M., and Charles E. Crede (1976). *Shock and Vibration Handbook*, 2nd ed., McGraw-Hill, New York.

International Organization for Standardization (1975). *Vibration and Shock—Vocabulary*, ISO 2041-1975.

Macinante, J. A., B. Dorien-Brown, J. L. Goldberg, N. H. Clark, R. A. Glazier, K. M. O'Toole (1967). A Vibration Study of the CSIRO 210-ft Radio Telescope, *Inst. Mech. Eng. London*, Mech. Eng. Sci. Monograph No. 6.

Manley, R. G. (1945). *Waveform Analysis*, Chapman and Hall, London.

Nolle, H. (1978). High Frequency Ground Vibration Measurement, *Shock Vibration Bull.*, **48**(4), 95–103.

Plunkett, R. (1982). Shock and Vibration Instrumentation, *Shock and Vibration Digest*, **14**(9), 3–5.

Rankine, W. J. M. (1874). *Songs and Fables*, in *A Book of Science Verse* pp. 149–151, selected by W. Eastwood, Macmillan, London, 1961.

3

Basic Principles of Vibration Isolation

The most infrequent usage of existing and developing shock and vibration technology is made in the area of design of vehicles, structures and machines; precisely where the most frequent usage should occur. The designer should be employing available technology to the extent required to avoid future problems in machine, vehicle or structural development and application. This utilization of shock and vibration technology could be looked upon as Vibration Insurance—insurance bought at a small premium that would cover later costly developmental programs or costly field failures.

R. L. ESHLEMAN (1973)

There is an extensive and growing literature on vibratory phenomena that are encountered in all fields of science and engineering, and a generous share of this literature is concerned with the control of unwanted and damaging vibration. The abundance of this literature is evident in the monthly publication, in the *Shock and Vibration Digest*,* of abstracts and reviews of the current literature on mechanical shock, vibration, acoustics, and related fields. In the year 1982 the *Shock and Vibration Digest* published more than 2600 abstracts of papers, articles, and documents, selected from journals, conference proceedings, trade magazines, and governmental and company reports published in many countries, which are regularly scanned. Reviews of the literature of general relevance to the subject area of this book have included Ward (1977, 42 refs.) on structural dynamic problems involving buildings, and Ungar et al. (1976, 72 refs.) on methods of estimating building vibration and reducing its severity.

Yet it seems that this fund of knowledge and information about vibration control is not finding general application in practice. Many of the architects and engineers who design buildings and the installations that will be used in them are either unaware of the possibility of vibration trouble, or optimistically believe they can attend later to any vibration problem that may arise. Some of those who do recognize the need for vibration isolation design mountings on rule-of-thumb formulas derived from oversimplified theory, as discussed in Chapter 6.

The failure to make adequate practical application of well-established principles and new theoretical insights is a result of ineffective communication between those who understand and develop theory and those who design hardware. Researchers go on studying vibration control problems of increasing complexity and describe the results with academic nicety at conferences and in the publications of learned societies. Teachers go on writing textbooks presenting the theory of vibration in generalized mathematical language that may be suitable in the academic environment but not in the design office. Meanwhile, the practitioner who lacks the time, the inclination, and perhaps the mathematical skill to understand theory presented in these ways simply ignores it.

The aim of this chapter is to stimulate practicing architects and engineers to recognize vibration as a factor that must be considered in the planning and design of their projects, and hence to invest in some vibration insurance as suggested by Eshleman in the opening quotation. Ways of doing this are summed up in the last section of the chapter. The book as a whole aims to bridge a part of the communication gap by explaining the physical basis of vibration isolation in a way that is consistent with rigorous theory but not burdened with mathematical derivations and proofs that can be found elsewhere.

---

* Published by the Shock and Vibration Information Center, United States Naval Research Laboratory, Washington, D.C.

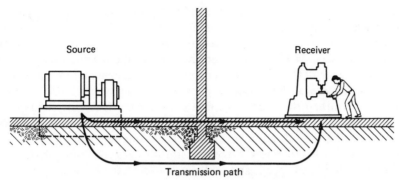

Figure 3.1.   Source, transmission path, and receiver of vibration.

The scope of vibration isolation is defined as follows with reference to Figure 3.1. Vibration is generated at a *source* of some kind and is transmitted through the ground and/or structure that forms the transmission path to a *receiver*—a term used here to include sensitive equipment, humans, or whatever is disturbed by the vibration.

The term *vibration isolation* implies that something is vibrating and that the vibration is objectionable for some reason. Vibration isolation is the action or treatment that eliminates the unwanted vibration or reduces it to an acceptable level. This may involve any or all of the following:

1. Action at the source to reduce the severity of the vibration transmitted from the source into its support or site.
2. Modification of the transmission path to reduce the severity of the vibration that is transmitted from source to receiver.
3. Action at the receiver to reduce its response to unwanted vibration of its support.

Although the insertion of vibration isolators under source or receiver could be regarded as modification of the transmission path, in this chapter we refer to vibration isolating mountings under the headings of action at source and receiver.

In the following sections we discuss these three areas of action in their application to existing unwanted vibration and to potentially troublesome vibration in proposed new installations.

ACTION AT THE SOURCE

When objectionable vibration occurs in a factory, laboratory, office, or residential building, common sense dictates that the source be identified so that any action possible may be taken to eliminate it or reduce its severity. Often the source is obviously a particular machine or process. Where there are several possible or suspect sources, the culprit may be identified by starting and stopping individual sources. Where this is not practicable, the troublesome vibration can be analyzed to determine its major frequency components, which are clues to the identification of the source (e.g., see Example 2.4).

Having identified the source we first consider the feasibility of reducing the vibration severity by modifying the source. Normally, the most practicable way of reducing the vibration severity of a rotating machine is by dynamic balancing, as discussed below. Other possibilities are changing the operating conditions of load and speed, and replacing the offending machine or process with one that generates less vibration. When more than one machine of a given type at a particular site generate excessive vibration, there is the possibility of arranging and operating them in such a way that the overall effect is minimized, as in the following case.

Example 3.1. An 800 hp, 180 rev/min gas compressor, having two horizontal cylinders arranged on opposite sides of the crankshaft and not in line, was installed directly on the concrete floor of a new brick factory which was built on piles. Severe vibration occurred and cracks appeared in a wall. Measurements showed that the entire factory oscillated about a vertical axis with a frequency of 3 Hz, and a displacement as high as 1.0 mm (p-p) was observed. On the recommendation of consulting engineers, the company purchased a modified crankshaft, but they did not install it because they could not permit the compressor to be shut down long enough for the changeover. They had become accustomed to the vibration, and evidently were more concerned about loss of production than loss of a few bricks.

Subsequently, when it became necessary to provide an additional compressor, one of the same kind was purchased. The two sets were fixed on a more massive foundation, and it was arranged that when both sets operated they would always be out of phase. With only one set operating the result was particularly beneficial because the total inertial mass of both sets opposed the unbalance.

If action to reduce the severity of the source is impracticable or ineffective, consideration should be given to the possibilities of changing the site of the source, or placing it on a seismic mounting. The factors to be considered in the siting and the seismic mounting of machinery are discussed below.

A decision about action at the source should, of course, be influenced by consideration of the action possible at the receiver. This is obviously advisable when there are many sources and relatively few vibration sensitive areas, as is common in factories.

In the design of new installations it is necessary to identify potential sources of vibration. In residential and office buildings these include the motors, compressors, blowers, pumps and other items of heating, ventilating and air-conditioning plant, the standby electricity generating set, and the elevator machinery. In factory buildings, apart from engines, compressors, and machine tools, there may be sources of mechanical shock or impact such as forging hammers, metal shears, and presses. In hospitals and other buildings where long-span suspended floors are used to meet the need for large open areas, persons walking on the floor may be a source of objectionable vibration. Failure to anticipate the need for vibration control may result in costly delay in placing new plant in service, as in the following instances.

Examples 3.2.   Vibratory conveyors on the upper levels of a new reinforced concrete building produced vibration so severe that operation of the plant could not be permitted until action was taken to reduce the vibration.

A vibration testing laboratory acquired an electromagnetic vibrator for testing aircraft equipment and assemblies to vibration-test specifications. It was realized only when first trials were made on the shop floor that the vibrator could not be operated until a seismic mounting had been provided.

Dynamic Balancing

The severity of vibration generated by rotating machinery can be reduced by dynamic balancing, which not only reduces the vibration transmitted into the surroundings, but also improves the running conditions, smoothness, and general "well-being" of the machinery itself.

The rotors may be balanced *in situ* or removed and balanced in a balancing machine. Large rotors are best balanced *in situ* not only because this eliminates the costly downtime involved in removing the rotor and transporting it to and from a balancing machine, but also because handling after balancing can introduce new unbalance. There are well-established techniques for balancing *in situ* and dynamic balancing services are commercially available. There are many examples of the use of *in situ* dynamic balancing to reduce objectionable vibration caused by rotating machines. The following are from case notes of acoustic consultant David Eden (personal communication).

Examples 3.3.   A ventilating fan located on top of a 10-story building was causing objectionable vibration. On one floor bottles on a certain shelf rattled, and on another floor water in an aquarium rippled when the fan was running. Dynamic balancing reduced the vibration to acceptable levels.

A fan was installed at the top of a tall flimsy steel structure and the vibration at the bearing due to its unbalance was excessive. The rotor was extremely difficult to balance, partly because of the fear that it might disintegrate, also because vibration amplitude readings were not repeatable on successive runs while the fan was running with a nominally constant condition of unbalance of the rotor. The reason for this became evident when it was noted that the steel casing of the fan was tearing away from the bearing pedestals; consequently, the stiffness of the fan was changing. After this was remedied by welding thick steel stiffeners between the fan pedestal and the casing, reproducible results were obtained, and the fan was satisfactorily balanced.

In the design of new installations, vibration should be recognized as one of the factors to be considered in the choice and specification of machinery. On considerations of vibration severity and balance quality, a purely rotational machine such as a turbine or a rotary compressor is preferable to a reciprocating machine. At present the relative severity of different types of machine is assessed mostly in the light of experience. With the development of internationally accepted methods of measuring and specifying machinery vibration, the severity of vibration sources will be described more and more in numerical terms. An acceptable level should be specified either in terms of the vibration of the machine as a whole, or the dynamic balance quality of its rotor, as discussed in Chapter 4.

The fact that the machine vibration severity is within the acceptable level for a machine of its type and application does not guarantee that the machine when installed will not cause objectionable vibration. The vibration of the completed installation will depend on where and how the machine is installed, so we now discuss the siting and seismic mounting of machinery.

Siting of Machinery in Buildings

Considering the building as a whole, the favorable location for vibration sources, particularly large machines, is at basement or ground level because the structural response increases with the height of the vibration source above the ground. A vibrating machine can excite horizontal vibration more readily at an upper than at a lower level. Also, because the upper stories are usually of lighter construction than the lower, a machine on an upper floor can more readily excite local vibration of floors and walls.

A major consideration is the need to avoid magnification of vibration by resonance, which occurs when an excitation frequency coincides with a free or natural vibration frequency of the structure. Machinery of the kinds normally installed in residential and office buildings is unlikely to cause resonance of a tall building in its fundamental rocking mode as a vertical cantilever because the natural frequency, typically below 5 Hz, is well below the usual range of machine vibration frequency. However, the possibility of resonance in higher modes of vibration of a tall building should not be overlooked. Resonance in the fundamental horizontal mode is more likely to occur in a building a few stories high, for example, an industrial building in which a vibratory conveyor is installed at an upper level. Local resonance of the part of the structure supporting a vibration source is a more common source of trouble. The method given in Chapter 8 for the design of seismic mountings for machinery on suspended floors shows how this resonance can be avoided.

The need to locate vibration sources where they will cause minimal structural response must be considered in conjunction with the desire to have them remote from vibration-sensitive areas. This of course demands a compromise not only insofar as the vibration factor clashes with the various other factors that determine layout, but also because a floor that is a desirable site for a machine because the floor would have small response to excitation applied directly by the machine, may also be a desirable site for a sensitive equipment because the floor would have small response also to indirect excitation from machinery elsewhere in the building.

Seismic Mounting of Machinery

If a particular machine is causing objectionable vibration that cannot be eliminated by modifying the operating conditions, by dynamic balancing, or by resiting the machine, there remains only the possibility of isolating the machine on a seismic mounting. If the machine is relatively small this represents a simple and economical solution of the problem. If the machine is large, its reinstallation on a seismic mounting will involve costly design, construction, and loss of output during downtime. Therefore before deciding to isolate the machine, consideration should be given to the possibilities, discussed later, of action at the receiver. Specifically, if the vibration is disturbing a particular item of sensitive equipment, isolation of that equipment may be preferable to isolation of the source of the vibration.

It is sometimes found that a source of objectionable vibration is already installed on isolators, but that these are ineffective because the isolators are misaligned or overloaded with the result that the load is no longer "floating" on the isolators, or the isolators are "bridged" by nonflexible piping or other connections. Obviously in such cases the load rating and number of isolators should be checked and if necessary modified, and any other installation faults rectified.

Example 3.4.   Two large Diesel electric standby sets, each mounted on eight rubber-in-shear isolators, were installed at the top of a newly completed high-rise building. Operation of the sets caused objectionable vibration in tenanted accommodation in the levels immediately below the plant room, and in the plant room itself.

An inspection by vibration consultants showed that eight of the 16 isolators were "grounded." The bolt head under the inner part that should be "floating" was bearing solidly on the floor plinth, as a result of misalignment and/or overloading. No provision had been made for convenient removal and replacement of individual isolators. The obvious short-term remedy was applied: a pneumatic drill was used to cut clearance in the concrete under the boltheads. This reduced the vibration to an acceptable level but, of course, left the misalignment and overloading to be remedied.

In the design of new installations, even if care is taken in the specification of the machinery and its siting, some machines will require seismic mounting. The decision whether or not to provide a seismic mounting for a particular machine is not always easy. A background of information on vibration criteria is given in Chapter 4 to help those having to make this decision. The types of seismic mounting are described in Chapter 5, and the design of source mountings is discussed in Chapter 8.

MODIFICATION OF TRANSMISSION PATH

Vibratory energy travels from a source to a receiver through whatever intervenes—the soil, sand, clay, rock or other material that makes up the "ground," or the reinforced concrete, steel, masonry, or other material of a building. Much has been written about the transmission of vibration over long distances from earthquakes and large explosions, and over short distances from quarry blasting, machinery, and other sources. Before considering ways of modifying the path to reduce transmission, we briefly note the following facts about the transmission of vibration through ground and structures. The reader who wants more should see Barkan (1962) and Richart et al. (1970) on ground vibration, and Biggs (1964) on structural vibration.

Transmission Through the Ground

In the vicinity of an impulse, such as that caused by blasting, pile driving, or a dropped weight, the ground responds with a transient vibration having characteristics that are determined by those of the source, and by the elastic and damping properties of the ground and the size and shape of any structures on the site where the transient is observed.

For site testing purposes a steady-state excitation can be imparted to the ground by using a mechanical oscillator in the form of a pair of counter-rotating unbalanced flywheels supported on a baseplate. The resultant of the centrifugal forces associated with the mass unbalance is a vertical periodic force, which is varied in frequency by varying the rotational frequency of the flywheels. It is an experimental fact that the rotational (excitation) frequency can be adjusted to produce ground resonance, which occurs when the excitation frequency is equal to a natural frequency of the particular site.

The predominant frequency of the transient response, and the lowest resonance frequency with steady-state excitation, are sometimes called the *site natural frequency.* This term should be used with care because this frequency is not an inherent property of the site. The natural frequency when loaded by a structure may be significantly different from that of the unloaded site. This is important in relation to the design of the foundation for a large engine or machine that is to be installed on a block formed directly in the ground.

The amplitude of vibration through ground having uniform properties decreases with distance from the source. Theoreticians have shown that even in a perfectly elastic medium the amplitude must decrease with distance from the source. In real earth materials the rate of decrease in amplitude with distance from the source is greater because of the energy dissipated in material damping.

One might expect that transmission through the ground could be blocked by a break or gap in the transmission path. There has been some interest in the use of a trench or void but the method seems to be unattractive and cumbersome. In effect, a trench or void merely lengthens the path, for the energy travels under the break. Richart et al. (1970, Chap. 8) give examples of the use of narrow trenches, ranging from 4.9 to 13.7 m deep, and describe their own experimental investigations of trench and sheet wall barriers. They make design recommendations based on the experimental results, and draw attention to the practical limitations of the method, not the least of which is the need to keep trenches open for depths of practical concern, say 6–15 m.

Transmission Through Structures

When an impulse is applied somewhere in a building, transient responses occur in all the beams, columns, floors, and walls. The amplitudes and frequencies of these transients are determined by the natural frequencies and damping of the individual members traversed. The response of one part of the structure may be markedly different from that of another.

When a source of steady-state excitation begins to act, each part of the structure makes a "starting transient" similar to that caused by an impulse. These transients decay after a few cycles, and thereafter the vibration everywhere in the structure has the same frequency as that of the source of the excitation. The magnitude of the response is amplified by resonance of any part of the structure whose natural frequency coincides with the excitation frequency.

The energy available to produce a response of a part of the structure decreases with the distance from the source. The rate of attenuation with distance depends on the damping properties of the materials traversed and the type of construction. The damping is greater in masonry and rivetted steel structures than in reinforced concrete and welded steel.

Structural Resonances

We have noted that the vibration along the transmission path is magnified by resonance and reduced by damping. It follows that the design policy for vibration control must be to avoid resonances and to incorporate as much damping as practicable.

The most troublesome resonance usually occurs at the beginning of the transmission path, in the structure supporting the source. This structure may be a foundation block set directly in the ground, a raised platform or pedestal, or a suspended floor. The amplitude of the alternating force transmitted through the supporting structure to other parts of the building is greatest at resonance; therefore, machinery installations should be designed to avoid operation at resonance. A typical recommendation is that a resonance frequency should not be within 20% of the shaft rotational frequency.

This requirement is taken most seriously in the design of structures for supporting large rotating machines such as turboalternators, particularly when the support is in the form of a relatively stiff table or platform supported on columns. The open nature of this arrangement is attractive because auxiliary equipment can be placed in the space immediately under the platform. The structure may be designed to be "low tuned" so that the fundamental natural frequency in a particular mode is lower than the operating frequency of the machine, or "high tuned" so that the fundamental frequency is above the operating frequency. Wilson (1974) reviews the relative merits of low and high tuned supporting structures, the methods of analysis of their dynamic behavior, and the design criteria.

If the machine is fixed without isolators to the structural support, the resonance frequency in a particular mode of vibration is that of the support as loaded by the machine. If the machine is seismically mounted, the machine and isolators constitute a second mass–spring system coupled to the support which itself behaves as a mass–spring system. Consequently there are now at least two resonance frequencies. The implications of this for the designer of the mounting are discussed in Chapter 8.

It is important to anticipate the possibility of resonance and to design to avoid it, because if a serious resonance occurs in the completed building usually it is impracticable to make large enough changes in the mass and/or stiffness to "detune" the system.

Dynamic Absorbers

Serious vibration of a structure or a part of it can sometimes be suppressed by a dynamic absorber, which is a secondary mass–spring system that is tuned to absorb energy from the primary system to which it is attached. The secondary mass is usually small in comparison with the primary mass. The motion of the secondary mass is opposite in phase to that of the primary mass, and therefore imparts a reaction that opposes the motion of the primary mass (Den Hartog, 1956, p. 93). Dynamic absorbers warrant consideration in practice only when the vibration to be suppressed is in a well defined direction and of constant frequency. The following is an example of the use of dynamic absorbers (Goldberg et al., 1980) to suppress severe vibration of the reinforced concrete roof of an aircraft engine test cell.

Example 3.5. Turbopropeller engines are tested to 4500 hp in a test cell 53 × 7 m, and 7 m high. The roof is of slab and beam construction; a typical slab over the test area is shown in Fig. 3.2. The slab thickness is 160 mm at wall boundaries and increases to 230 mm at the centerline.

From measurements made with a piezoelectric pressure transducer in the test space, the spectrum of pressure versus frequency showed a strong component at 68 Hz, corresponding to pulses from the four-bladed propeller rotating at 1020 rev/min.

Measurements of the roof slab vibration were made using six accelerometers together with filter amplifiers, a high speed chart recorder, and a multichannel instrumentation tape recorder. The chart record showed instantaneous values of vibration amplitude and phase, enabling the mode shape to be determined. Tape records were subsequently analyzed for frequency and average velocity values, using a real time analyzer. The results showed a remarkably pure mode shape (Fig. 3.2) at the 68 Hz frequency. At the antinodes the velocity amplitude at the engine power of 4000 hp was 24 mm/s, which, according to ISO 4866 (see Chapter 4), is in the zone of possible damage. Adjacent slabs showed similar modal behavior but lower amplitude.

The method of treatment involved placing a vibration absorber, consisting of a mass on rubber isolators, over an antinode to withdraw a significant part of the kinetic energy of the roof vibration. Some of the energy is dissipated in internal friction in the isolators, and some is transferred to the vibration of the absorber mass. The 235-kg absorber mass, 0.6 m in diameter and 0.4 m high, made up of eight concrete disks, rests on three isolators each consisting of a commercially available ribbed neoprene material 0.2 m in diameter, cemented between machined steel disks.

The degree of mode suppression achieved is shown in Fig. 3.3. With one absorber, there was partial suppression, and with two absorbers no point on

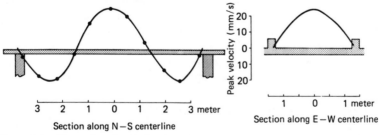

Figure 3.2. Flexural mode of vibration of concrete roof slab (Goldberg et al., 1980).

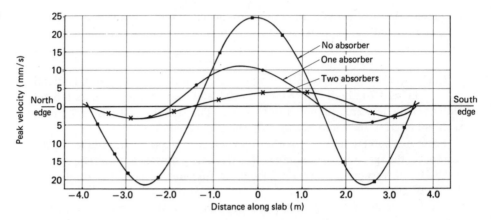

Figure 3.3. Mode suppression using one and two dynamic absorbers (Goldberg et al., 1980).

the slab had a velocity amplitude greater than 4 mm/s, as compared with 24 mm/s at the antinodes without the absorbers.

With the current popularity of very tall buildings and structures, the dynamic absorber is attracting attention as a means of controlling horizontal oscillation caused by wind gusting. Waller (1971, p. 320) suggests the use of a dynamic absorber in the form of a block of concrete on rubber bearings, which could be situated in a plant room at the top of the building. For the Centrepoint Tower in Sydney, Australia (Wargon, 1973), the water tank that supplies the fire sprinklers functions also as a dynamic absorber. The tank, which encircles the tower, is suspended on cables and contacts the tower through shock absorbers. The suspended system is tuned to dissipate the energy of wind-excited vibration of the tower.

Damping

The vibratory energy available to cause disturbance at the receiver can be reduced by increasing the damping in the transmission path. The internal damping inherent in most structural materials is relatively small and, until structural materials of high damping capacity become available as a viable alternative to the materials now used, the damping can be increased significantly only by designing separate damping elements into the transmission path.

Damping materials may be inserted at interfaces, for example, between the runway and the supports of a traveling crane to reduce the transmission of shock and vibration from the crane to machine tools in the vicinity. In concrete floors in factories and laboratories, materials such as cork and bitumen can be used to form the breaks necessary for crack control and at the same time reduce vibration transmission through the floor.

Methods of construction are being developed in which a "sandwich" layer of damping material is formed within a beam or slab so that vibratory energy is dissipated by the shear stressing of the damping layer. This method has been used to damp vibration in the beams and slabs supporting a railway track (Grootenhuis, 1968).

The results of the investigations referred to in Chapter 8 on the vibration isolation of machinery on suspended floors clearly show that transmission through the floor to other parts of the building decreases with increased damping of the floor. Unfortunately, it seems that materials and methods of construction are not yet available to take advantage of this fact by providing highly damped floors for plant rooms in buildings.

On the other hand, there have been some interesting developments in the damping of very tall structures to reduce their response to wind excitation. One example is the Drax power station in England (Waller, 1971, pp. 312–319). An outer shell or wind shield 260 m high and 26 m diameter contains three flues, each in 11 sections. Each group of three sections is supported on a platform that rests on laminated steel and neoprene bearings designed to function as dampers as well as expansion bearings. Another example is the damping of the 110-story twin towers at the World Trade Center in New York (Anon., 1971) in which viscoelastic shear dampers are used to limit wind induced vibration. At each floor there are 100 dampers connecting floor trusses to building frame.

Nakra (1976) gives a general review of the literature on vibration control with viscoelastic materials. Nelson (1977) reviews three methods for increasing damping: the design of structural joints and interfaces to promote damping; the use of layers of viscoelastic materials; and the use of discrete dampers.

ACTION AT THE RECEIVER

There are two main possibilities of action involving sensitive equipment that is receiving troublesome vibration through its support: the transfer of the equipment to a less-disturbed site, and the provision of a seismic mounting for the equipment.

In principle there is another possibility: the modification of the equipment to make it less sensitive to vibration. The theoretical basis and physical nature of the modification required will become evident in Chapter 9. In practice, however, modification of equipment to reduce its vibration sensitivity is likely to be feasible only with experimental equipment in research laboratories, and with prototypes of sensitive equipment in the developmental stage before manufacture.

In proposing new installations the potential users of sensitive equipment should consider vibration sensitivity as one of the factors influencing their choice of equipment. Unfortunately, at the present time they seldom can do so because few suppliers offer useful information about the vibration sensitivity of their products. All that can be done by the architects and engineers responsible for planning equipment layout and installations is to identify the equipment that is likely to be disturbed by site vibration, so they can give particular attention to its siting and provide seismic mountings where necessary.

Siting of Sensitive Equipment

The siting of sensitive equipment is of primary importance. The following is an example of the trouble that can result when a sensitive equipment is installed on an unsuitable site.

Example 3.6.   A large and costly camera for photolithography and map copying was installed on the first floor of an old timber building. The first-floor site was adopted for convenience of work flow and because the basement was occupied by printing machinery, although it was known that these cameras are normally installed in basement areas and remote from sources of vibration, to ensure good reproduction of fine linework. The 7-tonne camera, which was 9 m long, 3 m wide at the plate end, and 3 m high, was installed on rubber isolators of the type normally used at basement sites. The camera was unusable.

An attempt was made to overcome the trouble by stiffening the suspended timber floor supporting the camera, with an arrangement of structural steelwork and concrete. This involved costly design and construction, and continued loss of production, but proved unsuccessful.

At this stage the owners arranged for a vibration investigation, which showed that the predominant vibration was in the horizontal direction at right angles to the optical axis of the camera with a frequency of 3 Hz. This frequency was measured at a number of places throughout the building. As no source of this frequency operated in the building or in the neighborhood, the frequency of 3 Hz was assumed to be the natural frequency of horizontal vibration of the whole building. This would explain why the floor stiffening was ineffective: the floor, now stiffer, was still vibrating horizontally at 3 Hz. The problem was solved by providing a seismic mounting on helical springs (see Fig. 5.6) designed in consultation with the author (Macinante, 1961).

The author can recall many other instances of the loss of production, and the annoyance and frustration of the operators of sensitive equipment, that have resulted from the installation of sensitive equipment at unfavorable sites. The following are typical of cases where sensitive equipment was allocated an unfavorable site because considerations of work flow took precedence, or because a suitable site was not readily available, or simply because those concerned did not appreciate the importance of good siting. In all these cases the operators of the equipment were obliged to devote much time and effort to the vibration isolation of their instruments.

Examples 3.7. In a textile factory, balances and other instruments in the physical testing laboratory on the fourth floor were seriously disturbed by vibration from a large fan on the floor below. Removal or isolation of the fan was not feasible, and the dynamic balance of the fan could not be maintained because the blades became coated with airborne matter from the manufacturing process, and needed regular cleaning.

In a large mechanized foundry a sensitive balance was in use on a small table on the suspended timber floor of the works laboratory, about 100 m from a "jolter" that separated newly cast baths from their molds. The balance vibrated severely every time the jolter operated. The balance technician was expected to weigh microanalytical samples during a brief interval in each cycle of operation of the jolter.

In a factory mass-producing small packaged goods, two sensitive automatic weighing machines were installed on a suspended timber floor on which manufacturing machinery was operating. Although the vibration prevented the machines from operating properly, the management insisted that the weighing machines remain in their positions.

In a number of instances, electron microscopes, which on high magnification are highly sensitive to vibration, have been set up in very unfavorable sites: one shared a concrete floor with a resonance tester for large test objects, and another was on a suspended floor immediately over the laboratory air-conditioning plant.

In a large engineering works, because of the disturbance of sensitive equipment used for the physical and metallurgical testing, a new and well-constructed works laboratory was built at a site well removed from the heavy processing and manufacturing activities. Unfortunately, this was not enough, for the vibration isolation of equipment within the laboratory itself was unsatisfactory: microscopes, balances, and hardness testing machines were disturbed by the operation of machines used for preparing test specimens, and by physical testing machines such as tensile and impact testers.

As a general rule, the disturbance of sensitive equipment decreases with distance from the source. However, exceptions occur because the response of a floor, wall, or other part of a building, on which sensitive equipment may be installed, depends on its dynamic characteristics as well as on the energy input. For example, suppose that an instrument that is being disturbed by vibration from a particular machine is moved to a new site farther from the machine but one that happens to be on an unsupported area of a suspended floor. At the new position the energy input is less but the dynamic magnification may be greater and hence the disturbance greater.

It is worth emphasizing that sensitive equipment that must unavoidably be installed on a suspended floor is preferably located near a wall or on a part of the floor that is supported by columns or beams, because the vertical vibration of floor areas over or near the supports is always less than that at positions away from the supports.

A suspended floor is responsive not only to vibration received from sources elsewhere in the building but also to persons walking on the floor itself. This is particularly significant for long span reinforced concrete floors and, of course, timber floors. It should not, but evidently does, need saying (e.g., two cases mentioned in Examples 3.7) that a suspended timber floor is an entirely unsuitable support for sensitive equipment.

Another factor to remember in the siting of sensitive equipment is that the horizontal (swaying) oscillation of a building in its fundamental mode, caused by wind gusting or ground vibration increases with height above the ground (e.g., see Example 2.3). A seismic mounting for an installation on an upper floor may be worse than useless because the low-frequency horizontal motion, typically less than 5 Hz, may induce a large horizontal response of the seismic mass. This has recently been shown (Dorien-Brown, personal communication) to be major consideration in the siting of electron microscopes of the kind in which there is a seismic suspension within the equipment to isolate the specimen table from vibration of the base or frame of the equipment.

Laboratory benches for supporting sensitive instruments such as chemical balances should not be of lightweight wooden construction standing on the floor. Preferably the bench support should be of concrete construction, bolted to or integrally poured with the floor, and situated near walls or over columns or primary beams. If considered necessary, a seismically mounted slab can be provided on the bench at working height (Macinante and Waldersee, 1964).

The siting of a large and important equipment in an existing building is best decided on the results of a vibration measurement survey of the available alternative sites. The measurements preferably should determine both the "total" vibration and the frequency spectrum (Chapter 2) because the response of the equipment will depend on the frequency as well as the amplitude of the vibration at its site (Chapter 4, Sensitive Equipment; Chapter 9).

Seismic Mounting of Sensitive Equipment

When a sensitive equipment is being disturbed by vibration often the first thought is to put it on a seismic mounting. A decision to do so should not be made before considering the feasibility and relative merits of the alternative possible lines of action that have been discussed earlier in this chapter. This is particularly important when the decision relates to a large item whose reinstallation on a seismic mounting would entail costly design and construction, and loss of production during downtime.

If there is objectionable vibration of a sensitive equipment that is already installed on isolators, perhaps the seismic mass is not "floating" (see Example 3.8). The isolators may be misaligned, overloaded, or bridged by stiff piping or other connections. If correction of obvious faults does not remedy the trouble, and if it is not practicable to reduce the severity of the vibration received or to relocate the sensitive equipment, then a redesign of the seismic mounting is necessary (see Chapter 5, Design Guide).

Example 3.8.   In the first floor balance room of a newly constructed chemical laboratory a concrete floor, with integral pillars for the balances, was supported on rubber isolators, with an independently supported timber floor over it, on which the operators could walk without disturbing the "floating" floor (see Macinante and Waldersee, 1964, Fig. 14).

The result was completely unsatisfactory, for reasons that were not discovered because no provision had been made for access to the isolators for inspection or adjustment. The isolators may have been misaligned, or too stiff, or not stiff enough and hence overloaded; or there may have been contact where there should have been clearance between the floating floor and the adjacent structure. Whatever the explanation there was no alternative but to grout the floor solid and provide isolation between the tops of the pillars and the balances.

In planning the installation of equipment for a new building some items will require seismic mounting even though favorably sited in relation to vibration sources and structural support. In order to decide whether a particular item should be seismically mounted, information is needed on the acceptable level of site vibration for that item. The suppliers of sensitive equipment should be asked for advice on the environmental vibration level that their equipment can tolerate, and for recommendations about siting and vibration isolation based on their experience with the particular equipment. Some guidance can be obtained also from published information referred to in Chapter 4. The types of seismic mounting are described in Chapter 5, and the design of mountings for sensitive equipment is discussed in Chapter 9.

Whole-Building Isolation

When it is desired to isolate the occupants and/or equipment in a building from severe vibration generated outside the building, there is the possibility of isolating the whole building. This first attracted attention in the construction of buildings over railway lines in the 1920s in United States, where it became normal practice to support the columns on lead–asbestos pads.

A more effective method of whole-building isolation was introduced in the 1960s when R. A. Waller used natural rubber isolators of the kind developed as bridge bearings (see Figure 5.9) to isolate a four-story block of flats over St. James' Park underground railway station in London. The theoretical and practical considerations involved are described by Waller (1969).

Since that time the method has been used in England to isolate a medical research laboratory, a luxury hotel, a supermarket, a concert hall, and other buildings constructed over or near railway lines. In Australia, the new Law

Courts and the new Theatre Royal in Sydney are isolated in this way. The basic considerations in the design, construction and maintenance of buildings on elastomeric isolators are outlined in a Draft for Development issued by the British Standards Institution (1975).

Ungar and Kurzweil (1980) review the literature relating to the transmission of noise and vibration from subway trains to nearby buildings. They describe and discuss methods of reducing ground borne noise: these relate to the use of resilient elements in vehicle wheel and suspension systems and in track support systems, the design of tunnels, trenches and barriers, and the use of resilient elements to isolate buildings and parts of buildings.

VIBRATION INSURANCE

Afterthought or palliative action to provide or improve vibration isolation when construction to the original design is well advanced or complete is usually difficult and costly, and the quality of vibration isolation achievable may be limited by constraints of space and floor loading. The cost penalty is that of the redesign and reconstruction, and that resulting from delay in the occupancy of a building and in the use of machinery and equipment.

In the design of aircraft and suspension bridges, vibration is recognized as a vital factor because of bitter experience of catastrophic dynamic failures. In the design of buildings and machinery installations, neglect of the vibration factor seldom results in a spectacular failure; usually the penalty is annoying and costly delay in the completion and acceptance of the work.

How much better to anticipate vibration trouble or, in the words of Eshleman at the beginning of this chapter, to take out some vibration insurance. In the light of over 30 years experience of vibration control problems encountered by engineers and architects throughout Australia who discussed their problems with the author and colleagues at CSIRO Division of Applied Physics, the following vibration insurance policy is recommended.

Coordination of Planning

Establish and sustain effective coordination among the architects and the mechanical and structural engineers engaged in the design of a building. Without this coordination, decisions that one or other group might make within their own area of responsibility might frustrate or complicate the provisions for vibration control. For example, in designing a tall office or residential building the mechanical engineers determine the layout of the air-conditioning system, including the siting of the plant room, on considerations of cost and performance of the system. Meanwhile the architects, on entirely different considerations such as the intended activities of the occupants, the view, and remoteness from traffic noise, determine the arrangement of the accommodation. Surprisingly often, when coordination has been lacking it has not been realized until the building has been nearing completion that the occupants of prestige or elite accommodation would be disturbed by plant operating immediately overhead.

Examples 3.9. In one building high-velocity fans in the 14th floor plant room were making objectionable noise and vibration through the ceiling of the 13th floor suite, which was being prepared for a foreign embassy.

When the managing director of a company moved into his rooms on the top floor of a new building, he was annoyed by vibration from compressors installed on the roof.

Two large reciprocating air compressors on the 15th floor of a new building caused intolerable noise and vibration on the 13th and 14th floors. In this case the possiblity of providing effective isolation was limited because inertia blocks could not be used without exceeding the allowable floor loading.

When a laboratory or other building is being designed for users who will have special requirements relating to the installation and isolation of sensitive equipment, the architects and engineers must maintain effective communication not only among themselves but also with the intending users or their representative; otherwise, the special requirements may be overlooked or misinterpreted as the design develops.

Example 3.10. At an early stage in the planning of a new research laboratory, a scientist requested that independent pillars be provided in a basement room for microbalances. At a later stage and without reference to the scientist, it was decided to locate the air-conditioning plant in the basement, and the balance room overhead. On visiting the site some time later, the scientist was astonished to find that the pillars, which normally are raised to bench height through an on-grade floor, extended to a height of about 6 m where they supported a "floating" floor for the balance room. Worse still, the air-conditioning machinery was to be installed near the base of the pillars. This arrangement involving tall pillars was, of course, worse than useless because the pillars vibrated continuously and magnified the vibration, as was demonstrated by vibration measurements. There was no alternative to grouting solid the clearance space between the floating floor and the adjacent floor, and devising a balance-isolating bench above the floor.

Type of Construction

Consider vibration when deciding the type of construction and height of the building. For special-purpose buildings such as research and science teaching laboratories, in which sensitive equipment will be distributed over most of the occupied area, the vibration factor demands serious consideration when deciding the shape, height, and materials of construction of the building.

Although it is well known that tall buildings of modern lightweight construction are very responsive to vibration, financial constraints sometimes force the adoption of buildings of this kind for research and teaching laboratories. This may result in general dissatisfaction among those who subsequently use the sensitive equipment, unless very diligent attention is given to the specification and layout of machinery and equipment and the provision of seismic mountings, to compensate for the "liveliness" of the building.

Vibration was one of the major factors influencing the decision about the type of building and the materials of construction for the recently completed National Measurement Laboratory at Lindfield, N.S.W., Australia. The very high sensitivity of the techniques and instruments used in the Laboratory demands strictly controlled environmental conditions of temperature, humidity, and ambient vibration. A two-story design was adopted with load-bearing brick walls and reinforced concrete floors (Australia Parliamentary Standing Committee on Public Works, 1970).

Even in buildings in which there is no intention to use sensitive equipment, the vibration "liveliness" of modern types of construction may be troublesome to the occupants particularly in the vicinity of the plant room. At present, one can only anticipate the further development of high-damping materials and of methods of construction that will attenuate the vibration more effectively.

If the building is to be very tall and slender the possibility of motion sickness of humans occupying the uppermost levels when the building is subjected to wind gusting deserves consideration (Hansen et al., 1973). Some examples of dampers and dynamic absorbers to control the vibration of tall structures are given above.

Choice of Machinery

Select or specify machinery that generates minimum vibration. Although the primary aim is to select machinery that will satisfy all the functional and performance requirements with minimal capital outlay and running cost, the vibration and noise factor should be considered also, particularly for residential and office buildings, hospitals, laboratories, libraries, concert halls, and the like. The requirement of minimum vibration and noise should not be overruled in the interests of economy, on the assumption that vibration isolating mountings and acoustical treatments can be devised later if necessary to cope with excessive vibration and noise.

An increasing number of users and manufacturers of machinery are adopting the recommendations of the International Organization for Standardization (ISO) relating to the balance quality and the vibration severity of machinery

(Chapter 4). By using the appropriate ISO or national specifications a level of acceptable vibration severity can be specified quantitatively.

An important item of equipment in modern buildings is the standby electricity generating plant, which at the present time usually has a reciprocating engine as prime mover. When installed on an upper floor of a tall building the standby set can cause severe vibration and annoyance unless it is effectively isolated (e.g., see Example 3.4). The increasing availability of gas turbines suitable for use as prime movers in buildings is pleasing because, as with any machinery of the purely rotational kind, they are more easily and effectively dynamically balanced and vibration isolated than reciprocating machinery.

Layout of Machinery and Equipment

Arrange favorable siting of machinery and sensitive equipment. This involves consideration of the relative locations of the sources and receivers of vibration, and the suitability of their structural supports, as discussed earlier.

Although those engaged in the design of a building usually choose or specify the machines that will become the sources of vibration, they would normally have no role in the selection of the sensitive equipment that they will be required to install in the building. They can only seek information about the sensitivity of the equipment so that, in consultation with the intending user, they can decide whether special attention should be given to siting or seismic mounting.

Vibration sensitive equipment and processes should be positioned as far from vibration sources as is practicable and compatible with the other factors that determine the layout, and which understandably may take precedence. For example, in a factory the necessary sequence of the manufacturing operations may place a precision machine tool or measuring instrument close to an operation or machine that is a vibration source; on a railway platform a ticket selling computer cannot be sited far from the tracks. In residential and office buildings the siting of the plant room in relation to a luxury or elite suite or other vibration sensitive area needs special care (e.g., see Examples 3.9). In hospitals and laboratories also, the plant room should be well separated from vibration sensitive areas.

Vertical separation of source and receiver is not likely to be effective in a tall building. The upper floors are not suitable for either sources or receivers because the responsiveness of the building to sources both inside and outside the building increases with height above the ground. Some decades ago the plant room normally occupied the basement. Today, unfortunately, the common practice is to locate the plant room on an upper floor or at the top of a tall building, because the basement is required for car parking or other purposes. The siting of machinery or sensitive equipment on upper floors makes severe demands on the quality of vibration isolation that is necessary.

When a site has been decided for a vibration source, the design of the structural support should be checked and if necessary modified to avert the possibility of resonance. This is most important if the machine is to be installed on a raised platform on columns, or on a suspended floor (Chapter 8). Similarly, the local structure at the site for a sensitive equipment should be designed to have minimal vibration even if the equipment is to be installed on a seismic mounting, because a mounting can achieve only a certain attenuation; the smaller the input through the support, the smaller the disturbance of the equipment.

Seismic Mountings

Decide what seismic mountings should be provided and ensure that any special requirements affecting structural form and load carrying capacity are satisfied. Even if the best compromise has been made in the choice and layout of the machinery and the sensitive equipment, certain machines and equipment normally require isolating mountings. Machinery that is known to vibrate severely and equipment known to be very sensitive to vibration obviously require seismic mountings. In doubtful cases, one must decide whether the additional cost incurred by seismic mounting is justified as an insurance against the losses that can be envisaged if the mounting is not provided. The discussion of vibration criteria in Chapter 4 provides some background to this decision.

The design of mountings should begin early so that any requirements affecting the shape and stiffness of the supporting structure may be incorporated in the structural design, and provision made for any pits necessary for below-floor mountings, "housekeeping" pads for above-floor mountings, and any other details affecting the structural design and construction (see Chaper 5, Design Guide.)

Inspection

Ensure that the seismic mass, as constructed, is "floating" with the intended clearances, on isolators that are correctly aligned and loaded. In the inspection and supervision that normally accompany the progress of the construction of a building, special attention should be given to the seismic mountings.

It is important, particularly when the isolators are out of sight under an inertia block, to ensure that the isolators are correctly loaded and aligned and that the specified provisions have been satisfied, such as protection from corrosion, oil, or other hostile environment.

Where a seismic block, or a section of "floating floor" is intended to be separated with a relatively small clearance (e.g., 50–100 mm) from the adjacent floor or structure, it is important to ensure that the clearance space is not filled with waste materials that are easily and conveniently shoveled or swept off the floor. The author was told of a case, understandably unpublished, in which workmen, thinking that a gap was only temporarily left, poured concrete into the clearance space and thereby "fossilized" the isolators.

Finally, when arrangements are being made for acceptance testing, the entire installation should be checked again to confirm that there is no bridging or other unintended connection between the "floating" and fixed parts before time and money are spent on acceptance tests involving operation of the plant and vibration measurements.

Acceptance Testing

Specify conditions for acceptance of the completed mounting. The tests and measurements that will be made to determine whether a seismic installation is acceptable should be clearly specified. Reliance on subjective judgment is unhelpful if a dispute arises. Not much more helpful is the stereotyped clause commonly found in building specifications: "the isolation efficiency for rotational plant shall be 90 per cent." This is meaningful only for a mass–spring system of impeccable behavior, vibrating purely vertically, for example; the kind found only in text books.

A more realistic specification concerns itself with the physical quantity that really matters: for a source mounting, the vibration that the machine causes in specified vibration-sensitive areas; and for a sensitive equipment mounting, the vibration of the critical parts of the equipment.

An acceptance test based on these criteria, which involve only the vibration transmitted by the mounting, is not necessarily a test of the mounting. For example, the vibration that a machine causes at a particular area elsewhere in the building is determined not only by the seismic mounting but also by the transmission path and the response of the structure where the measurement is made. The performance testing of the mounting involves some practical difficulties, which are discussed in Chapter 5, under the heading Performance Testing.

REFERENCES

Anon. (1971). Dampers Blunt the Wind's Force on Tall Buildings, *Architectural Record,* **150** (Sept.), 155–158.

Australia Parliamentary Standing Committee on Public Works (1970). *Report Relating to the Proposed Construction of National Standards Laboratories at Bradfield Park, New South Wales,* Parliamentary Paper No. 112, Commonwealth Government Printing Office, Canberra.

Barkan, D. D. (1962). *Dynamics of Bases and Foundations,* McGraw-Hill, New York.

Biggs, John M. (1964). *Introduction to Structural Dynamics,* McGraw-Hill, New York.

British Standards Institution (1975). *Vibration Isolation of Structures by Elastomeric Mountings,* Draft for Development DD 47.

Den Hartog, J. P. (1956). *Mechanical Vibrations*, 4th ed., McGraw-Hill, New York.

Eshleman, R. L. (1973). Vibration Insurance, *Shock Vibration Digest*, **5**(2), 1.

Goldberg, J. L., N. H. Clark, and B. H. Meldrum (1980). An Application of Tuned Mass Dampers to the Suppression of Severe Vibration in the Roof of an Aircraft Engine Test Cell, *Shock Vibration Bull.*, **50**(4), 59–68.

Grootenhuis, P. (1968). Sandwich Damping Treatment Applied to Concrete Structures, *Philos. Trans. R. Soc. London Ser. A*, **263**, 455–459.

Hansen, Robert J., John W. Reed, and Erik H. Vanmarcke (1973). Human Response to Wind-Induced Motion of Buildings, *Proc. Am. Soc. Civil Eng.*, **99**(ST7); *J. Struct. Div.*, **7**(9868), 1589–1605.

Macinante, J. A. (1961). Spring Mounting for a Large Camera, *Engineer* **212**, 1080–1081.

Macinante, J. A., and J. W. Waldersee (1964). The Vibration Isolation of Knife-Edge Balances, *J. Sci. Instrum.*, **41**, 1–6.

Nakra, B. C. (1976). Vibration Control with Viscoelastic Materials, *Shock Vibration Digest*, **8**(6), 3–12.

Nelson, F. C. (1977). Techniques for the Design of Highly Damped Structures, *Shock Vibration Digest*, **9**(7), 3–11.

Richart, F. E., Jr., J. R. Hall, Jr., and R. D. Woods (1970). *Vibrations of Soils and Foundations*, Prentice-Hall, Englewood Cliffs, New Jersey.

Ungar, Eric E., Clive L. Dym, and Robert W. White (1976). Prediction and Control of Vibrations in Buildings, *Shock Vibration Digest*, **8**(9), 13–24.

Ungar, E. E., and L. G. Kurzweil (1980). Means for the Reduction of Noise Transmitted from Subways to Nearby Buildings, *Shock Vibration Digest*, **12**(1), 5–12.

Waller, R. A. (1969). *Building on Springs*, Pergamon, London.

Waller, R. A. (1971). The Design of Tall Structures with Particular Reference to Vibration, *Proc. Inst. Civil Eng.* **48**, 303–323.

Ward, H. S. (1977). The Characteristics of Dynamic Loads and Response of Buildings, *Shock Vibration Digest*, **9**(8), 13–20.

Wargon, Alexander (1973). The Centrepoint Project, *Conf. on Planning and Design of Tall Buildings*, Sydney, Australia, Institution of Engineers, Australia, New Zealand Institution of Engineers, University of Sydney, and International Association for Bridge and Structural Engineering.

Wilson, R. R. (1974). Flexible Supports for Rotating Machinery, *Shock Vibration Digest*, **6**(6), 2–6.

4

Vibration Criteria

According to tape recordings made by the Royal Institute of Technology in Stockholm, birds of the family Picidae display an impressive tempo in their pecking. Thus, for example, the greater spotted woodpecker makes about 23 beak strokes per second, the lesser spotted woodpecker about 20, and the three-toed woodpecker about 14. An unmated woodpecker rattles off some 500–600 flourishes a day, while after mating the intensity falls off to 100–120 flourishes a day.

The reason why the woodpecker does not suffer from headaches is that it is specially equipped by nature to stand up to such extremely severe battering. Its beak is not rigidly joined to the frontal bone of the skull as in other birds, but is provided with a resilient pad that cushions the vibrations.

Compared to the woodpecker, man is poorly equipped to cope with vibrations. . . .

SVEN-ÅKE AXELSSON (1968, p. 9)

Certainly a human would not tolerate the kind of vibration that is ordinary and natural to the woodpecker. To have one's skull subjected to a succession of impacts at the rate of some 20 per second, and this repeated every 10 min or less is unthinkable.

Human sensitivity to vibration applied to particular parts of the body such as the head, hands, and feet, through vibrating headrests, handles, and pedals, is an important factor in the design of certain kinds of machinery, notably mobile earthworking and mining machines. In the design of buildings, whole-body vibration is the important consideration. We are concerned with vibration that is transmitted to the body as a whole through the feet when standing, the buttocks when seated, or the supporting area when reclining.

The surprisingly high sensitivity of humans to whole-body vibration raises problems of vibration control in residential buildings, A seemingly trivial level of vibration can cause annoyance and resentment, particularly when the vibration occurs intermittently and during the night when the level of other ambient vibration is very low. Even so, humans are not nearly as sensitive as some of the equipment used in the laboratories of research and teaching establishments and hospitals. The sensitivity of humans and of instruments is discussed quantitatively later in this chapter.

When complaints about disturbances of humans or their equipment are being investigated, consideration of the various kinds of remedial action discussed in Chapter 3 may indicate that isolation of the vibration sources is required, and presumably should have been provided in the original construction.

Although the need for vibration isolation may be obvious in hindsight, it may be far from obvious in foresight to those planning a new building and its installations, when neither the vibration nor its objectionable effects yet exist. Nevertheless, at this early stage it is necessary to decide, at least for each large item of machinery and equipment, whether a seismic mounting is necessary because, if it is, the load-carrying capacity and detail of the floor or other structural support must be designed at the outset to suit the installation.

In order to decide whether a seismic mounting should be provided for a particular machine we must first ask: What vibration will this machine generate in operation? Could the vibration be severe enough to damage the floors and walls of the plant room? What vibration will be transmitted to other parts of the building? Will it distract or annoy those working in office and laboratory areas or those relaxing or sleeping in residential areas?

The probability of the occurrence of any of these objectionable effects cannot be assessed without estimating the vibration that may occur and comparing this with the level of vibration that can damage structures, disturb instruments, and annoy humans. Investigations aimed at establishing these levels of acceptable or tolerable vibration have yielded an extensive literature. In this chapter we refer to only a small number of publications, selected to give a general idea of what level of vibration is and is not acceptable in given circumstances.

The broad aim in this chapter is to give some guidance and background to those who must decide, in advance of its installation, whether a particular machine or other equipment does, does not, or maybe needs a seismic mounting. If "maybe," then the decision must be made on the basis of a comparison of the additional cost that would be incurred by providing the mounting with the probable cost or other penalty of failure to provide the mounting.

MACHINERY VIBRATION SEVERITY

The major cause of vibration of any machine is dynamic unbalance of its moving parts. Designers and manufacturers aim to produce machines that are dynamically balanced to the highest quality that is practicable and profitable, in order to minimize both the alternating stressing of shafts, bearings, and frame of the machine, and the alternating force that is transmitted into the structure supporting the machine.

The vibration of the machine as a whole is not simply and directly determined by its dynamic unbalance. The dynamic unbalance of a rotor generates a vibration component having a frequency equal to the rotational frequency of the rotor—sometimes called the "once-per-revolution" component. Other sources of vibration within the machine contribute components having other frequencies; hence, there is need for criteria for the vibration of the machine as a whole.

Nevertheless, dynamic unbalance of the rotating parts of a machine is the most obvious, and commonly the most serious, source of vibration and hence should be minimized. In the following paragraphs we discuss criteria first for the dynamic balance quality of a rotor and then for the vibration severity of a machine as a whole.

Dynamic Balance Quality

The balance quality of a rotor is determined by the amount of its residual or uncorrected unbalance. Therefore, we must define "unbalance" before discussing balance quality in numerical terms.

We begin by considering the simplest rotor, a disk with two integral journals supported between bearings as in Fig. 4.1a. Because the thickness of a disk is small in comparison with its diameter we can consider the unbalance to be confined in a single plane, the mid-plane of the disk.

If the disk is of uniform density and has axial symmetry about the axis normal to the disk there can be no unbalanced centrifugal force when the disk is rotating about this axis. This same condition of perfect balance can be achieved even if the disk has neither uniform density nor axial symmetry, if the centrifugal forces acting radially on all the mass elements of the disk completely balance one another.

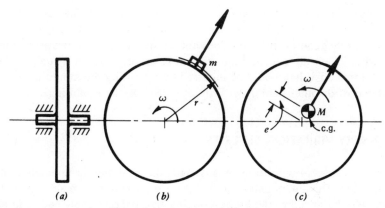

Figure 4.1.   Static unbalance of disk.

If this condition is not satisfied there is a resultant or unbalanced centrifugal force, which causes vibration of the bearings. There are two ways of describing the unbalance. It can be attributed to a "heavy spot" represented by a small mass m at radius r on an otherwise perfectly balanced disk as in Fig. 4.1b. This causes a centrifugal force $mr\omega^2$ where ω is the angular velocity of rotation of the disk. Alternatively, the unbalance can be considered to have displaced the center of gravity (c.g.) of the whole disk through a small distance e from the axis of rotation that is defined by the journals and bearings as shown in Fig. 4.1c. The unbalanced centrifugal force is that of the disk mass M rotating with radius e; this is $Me\omega^2$.

Since these are two ways of describing one and the same centrifugal force, we can equate the expressions:

$$mr\omega^2 \;=\; Me\omega^2$$

or

$$e \;=\; \frac{mr}{M}$$

This indicates that the unbalance can be expressed either as the mass radius product mr per unit mass of the disk, or as the displacement e of the c.g.

This condition of unbalance is called static unbalance because the "heavy spot" can be found without driving the disk. If the journals are placed on a pair of knife edges, truly horizontal and rigidly supported, and if the rolling friction is negligible, the disk will roll into the position where the heavy spot occupies the lowest position. Static unbalance is corrected by removing mass at the angular position of the heavy spot or adding mass opposite this position.

When the axial dimension of a rotor is not small in comparison with the diameter, the rotor could be considered to be made up of a large number of disks side by side, each having its own static unbalance as just discussed. The overall effect of the static unbalance in all the separate disks would be equivalent to having a heavy spot m_1 at a certain angular position somewhere near one end of the rotor and another heavy spot m_2 near the other end in a different angular position, as in Fig. 4.2.

Alternatively we can consider the unbalance of a rotor in terms of its principal axis of inertia. It is shown in textbooks on mechanics that, regardless of variation in shape and mass distribution along the rotor and assuming no elastic deflection (that is, a "rigid" rotor), there is a principal axis of inertia, conceptually fixed in the rotor, about which the moment of inertia is a minimum and about which the rotor would rotate without vibration if there were no constraint from the method of driving the rotor or of supporting it in bearings. If the shaft or journals are coaxial with this principal axis there will be no vibration of the bearings when the rotor is revolving.

When this ideal is not attained, we have the condition shown in Fig. 4.2. A rigid rotor is supported in bearings on journals whose axis is AA' the nominal axis or center line of the rotor. Because of unbalance of the rotor, represented schematically by the "heavy spots" m_1 and m_2, the principal axis of the rotor is some line OO' which is not coincident with AA'. The bearings and journals constrain the rotor to revolve about the axis AA'; therefore, each journal applies a force radially to its bearing. In most machines the consequent vibration is mainly horizontal because usually the horizontal stiffness of the bearing supports is less than the vertical.

In general, the c.g. of the rotor will not be on AA' but offset by a small amount, like that of the disk in Fig. 4.1c. Consequently, at the bearings there will be in-phase components of vibration resulting from the offset c.g. and antiphase components resulting from transverse oscillation of the rotor about its c.g.

The unbalance of any rigid rotor can be corrected by adding masses in any two planes such as those marked I and II in Fig. 4.2 (e.g., see Den Hartog, 1956, pp. 233–234). The dynamic balancing process, either in a balancing

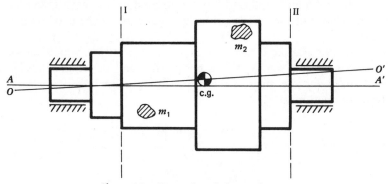

Figure 4.2.   Dynamic unbalance of rotor.

machine or *in situ*, determines the magnitude and angular position of the mass radius product that is required in each of the correction planes. The effect of applying the corrections is to bring the principal axis OO' into coincidence with the axis of rotation AA'. More information on the principles and practice of dynamic balancing can be found in Muster and Stadelbauer (1976).

There is always some residual unbalance. Not only does the balancing process have a limited sensitivity or capability of resolving the amount and angular position of the required correction, but also the balancing process is not necessarily taken to the limit of its capability because the nature of the rotor and its intended application may not warrant the extra cost of balancing to the highest possible quality.

In the context of vibration control, the question for the practitioner is: how can I specify for a potential source of vibration a balance quality that is adequate and reasonable, not exaggerated and uneconomical? The specification of unbalance tolerances has concerned manufacturers and users of rotating machinery for many decades and a variety of empirical formulas have been developed on the basis of experience to express the acceptable amount of unbalance of a particular class of rotor in terms of the mass and operating speed. A survey of American practice (Muster and Flores, 1969) covered rotors, shafts and wheels used in aircraft engines, electric motors and generators, centrifuges, compressors, turbines, and various other kinds of machinery.

With the object of establishing an internationally recognized method of specifying balance quality, representatives of the national standards bodies of some 20 member countries of ISO (the International Organization for Standardization) after meetings and discussion over a decade, have approved the method described in ISO 1940 (International Organization for Standardization, 1973).

The ISO document proposes a family of grades of balance quality which are intended to serve as a means of classification of balance quality, and thereby facilitate mutual understanding between manufacturer and user. A few examples of the many grades of balance quality defined in ISO 1940 are shown in Fig. 4.3 for reference in the following discussion. For each grade the acceptable unbalance of the rotor is given as a function of the maximum speed of rotation.

This method of grading is based on practical experience, which has shown that, for rotors of a given type, the permissible unbalance depends on the rotational speed and the mass of the rotor. The grade numbers are based on the following relationships:

$$e\omega = \text{const.}$$

or

$$\frac{mr\omega}{M} = \text{const.}$$

where e = displacement of c.g. (μm)

m = permissible unbalanced mass (g)

r = radius to mass m (mm)

M = mass of rotor (kg)

ω = $2\pi n/60$ = angular speed (rad/s)

n = rotational speed (rev/min)

The unbalance is expressed as the unbalance per unit mass in g·mm/kg or as the "center-of-gravity displacement e in micrometers." The following example indicates how these alternative expressions may be used.

Example 4.1. A balance quality grade of G6.3 is specified for a rotor of 40 kg mass and operating speed 3000 rev/min. What is the acceptable unbalance?

Referring to the line G6.3 in Fig. 4.3, for n = 3000 rev/min the ordinate is 20, which can be interpreted in two ways. The 20 can be taken as mr/M = 20 g·mm/kg, which means that the acceptable residual unbalance is 20

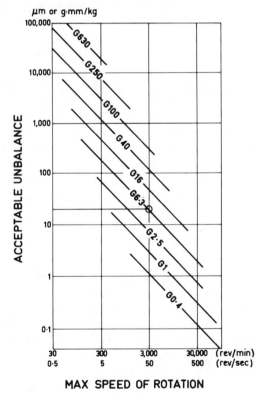

Figure 4.3. Balance quality grades (adapted from ISO, 1940–1973).

g·mm per unit mass (kg) of the rotor. Thus, for the 40 kg rotor, the acceptable residual unbalance is 800 g·mm.

Alternatively the 20 can be taken to indicate that the acceptable displacement (e) of the c.g. of the rotor is 20 μm (0.02 mm). Therefore for a 40-kg rotor the acceptable unbalance is $eM = 0.02 \times 40 = 0.8$ kg·mm, or 800 g·mm.

If the rotor is balanced by corrections made in two planes, the permissible residual unbalance is taken to be 400 g·mm in each plane if the c.g. of the rotor is located within the mid-third of the distance between the bearings and the correction planes are equidistant from the c.g. If these conditions are not satisfied the 800 g·mm is apportioned between the two correction planes in accordance with the mass distribution of the rotor.

ISO 1940 suggests suitable balance quality grades for various groups of rotor types. The grade suggested is not intended to serve as an acceptance specification for a particular group, but rather to indicate a grade that is neither a grossly deficient nor an exaggerated or unattainable quality requirement. The purchaser can specify the grade considered necessary on the basis of experience with the particular type of machine and its application.

Vibration Severity Rating

The fact that the rotor of a machine is balanced to the specified quality grade does not guarantee that the vibration of particular parts, or the machine as a whole, will be acceptable. The vibration that results from the residual uncorrected unbalance is influenced by the dynamic characteristics of the machine frame and the structure supporting the whole machine. Thus, the vibration of the bearings supporting the rotor will be influenced by the mass and stiffness of the bearing supports. The vibration of the machine frame and bedplate will depend on their stiffness and mass distribution. The vibration of the machine as a whole will be influenced by the stiffness and mass distribution of the floor, pedestal, or other structure supporting the machine.

In addition to the vibration at shaft rotational frequency resulting from unbalance, there will be components at other frequencies resulting from shaft misalignment, gear meshing, and imperfections of bearings and couplings, and from hydrodynamic, aerodynamic, and thermal effects.

Obviously criteria are required for acceptable vibration of machines independently of criteria for the balance quality of their rotors. The recognition of this need has sustained interest in this topic over some four decades since T. C. Rathbone, then chief engineer of a machinery insurance company, published his machinery vibration tolerance chart (Rathbone, 1939) as a guide for estimating the severity of machinery vibration.

In order to establish criteria it is necessary to decide what physical quantity (displacement, velocity, acceleration) and what measure of this quantity (peak, average, rms) should be used to characterize the intensity or severity of vibration. Other details to be decided are the places on the machine where the measurements should be made, and the characteristics (range and accuracy) that should be specified for the vibration measuring instrument. With regard to the testing of a machine to determine its vibration severity rating, the conditions of operation (load, speed, etc.) must be defined, as well as the manner of supporting the machine when making the measurements. Finally, numerical values must be decided for the levels of vibration that are acceptable for machines of the various classes.

Many different methods of measuring and describing vibration severity have been used by private companies and governmental establishments in various countries. Eshleman (1976) summarizes existing practices in relation to many classes of machinery including compressors, electric motors, fans, gear units, pumps, and turbomachinery.

With the object of establishing internationally acceptable criteria that would facilitate communication and understanding between the manufacturers and users of machinery, representatives of some 20 member countries of the International Organization for Standardization, after discussions extending over a decade, agreed to define "vibration severity" in terms of the rms value of the vibration velocity in ISO 2372 (International Organization for Standardization, 1974a). The vibration is measured at certain points and in defined directions, and the highest value is taken to be the "vibration severity." ISO 2372 defines 15 vibration severity ranges: the lowest for velocity up to 0.11 mm/s rms, and the highest to 71 mm/s rms as shown in the upper right part of Fig. 4.4. These ranges are defined for vibration measured at machine surfaces, such as bearing caps, and within the frequency range 10–1000 Hz and the speed range 10–200 rev/s.

ISO 2372 gives examples of the application of this scheme to four classes of machine: small, medium, large, and turbomachine. Ranges of vibration severity are suggested for quality judgment ranging from "good" to "impermissible." As an illustration, the ranges are shown in Fig. 4.4 for Class III, defined as "large prime movers and other large machines with rotating masses, mounted on rigid and heavy foundations which are relatively stiff in the direction of vibration measurement."

Unfortunately, the vibration severity of a machine measured in the prescribed way at the manufacturer's works may be significantly different from that measured by a user who has installed the machine on a support having dynamic characteristics different from those of the support used by the manufacturer.

In order to reduce the misunderstanding and disagreement that may arise in these circumstances, ISO 2372 points out that comparable vibration levels of machines under test are most readily achieved when the machines are "soft-mounted," so that the lowest natural frequency of the machine on the test mounting is less than one-fourth of the lowest excitation frequency. ISO 2372

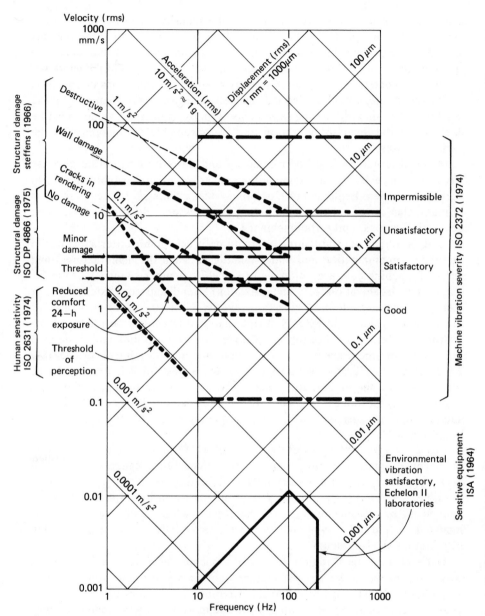

Figure 4.4. Vibration criteria for machinery, structures, sensitive equipment, and humans.

defines the conditions to be satisfied for a soft mounting. For machines too large to be soft-mounted for test purposes, the machine is of necessity tested *in situ* and the relevant details of the structural support should be described in the test report.

The classification scheme is available for machinery in general. Recommendations are being developed also for particular classes of machine. Two documents have been published: ISO 2373 (International Organization for Standardization, 1974b) for a certain class of electrical machine, and ISO 3945 (International Organization for Standardization, 1977) for large rotating machines operating at speeds from 10 to 200 rev/s.

An indirect method of setting a limit to the vibration acceptable at a machine installation is to specify the acceptable level of vibration that operation of the machine may cause at specified vibration-sensitive areas in the building, such as laboratory, office, and residential areas. Criteria of acceptable vibration for buildings, sensitive equipment, and humans are discussed in the following sections.

STRUCTURAL DAMAGE

When a residence is situated close to a factory, railway, or construction site, the householder becomes annoyed and resentful if vibration disturbs his or her rest or concentration, and also may fear that the vibration is damaging the property. Hostility increases greatly if cracking of ceiling plaster, wall rendering, or brickwork is discovered; damage that may well have existed unnoticed before the vibration began.

We know that severe vibration in the form of an earthquake can wreck buildings, and that even much less severe vibration from industrial sources such as blasting, pile driving, and heavy machinery can damage buildings in the vicinity. Yet, we know also from experience that floors and walls of factories and power houses can sustain perceptible vibration over periods of decades without damage resulting from the vibration. Where is the dividing line between damaging and safe vibration?

We need the answer for two main reasons: as a basis for deciding whether vibration from a particular source is likely to cause damage if the source is not isolated from the supporting structure, and as a basis for dealing fairly with complaints that vibration is causing or may cause damage.

In attempting to establish a scale showing numerical levels of vibration severity that are likely to cause various degrees of damage, we must first adopt some physical quantity as a measure of vibration severity. Steffens (1966) reviews in some detail various units that have been used as a "damage figure" or "strength" or "power" of vibration. We need not concern ourselves with these in the present context because in practice today it is generally accepted that the best indicator of structural damage potential is the vibration velocity.

The possibility of deriving damage criteria by calculation is virtually nil. A safe level of vibration is such that the stress caused by the vibration is less than the maximum stress that the material can withstand without cracking. Calculation of the maximum safe vibration would require knowledge of the maximum permissible stress and of the relationship between stress and vibration. We do not know the permissible stress, under dynamic conditions, for materials of the many kinds and qualities used in building construction. Also, we could not calculate with much confidence the dynamic stress caused by a given vibration, taking into account the distributed and usually nonuniform mass, stiffness, and damping of the structure, the inhomogeneity of the materials, and the possibility of resonance magnification.

A more reliable way of deriving damage criteria would be to vibrate a building under controlled conditions, measure the vibration of floors, walls, and other parts of the building and correlate the magnitude and duration of the vibration with damage observed to have been caused by that vibration. Testing involving intentional damage has been done with buildings intended for demolition to observe damage caused by transient vibration from blasting, as mentioned below, but not that caused by continuous vibration. As far as the author is aware, steady-state vibration testing of whole buildings has been done only at vibration levels well below those necessary to cause damage, the object being to determine modes of vibration and natural frequencies.

There has been some vibration testing of parts of buildings such as floors, walls, and ceilings to determine damage criteria. Data derived from these tests, and from investigations of vibration problems involving structural damage, are reviewed by Steffens (1966) who discusses the risk of vibration causing cracking of plaster, window glass, and brickwork. Steffens gives levels of vibration, shown on the upper left in Fig. 4.4, defining four degrees of damage: no damage, cracks in rendering, wall damage, and destructive. The lines defining these zones are not parallel to the grid lines of displacement, velocity, or acceleration because they are based on the "vibrar," a more complicated unit which is explained in Steffens' paper.

These same levels are given by the Great Britain Building Research Station (1970) as the most reliable information available for the estimation of structural damage from vibration. BRS states that there is no evidence that fatigue is significant; usually there is a large factor of safety against fatigue failure of steel, reinforced, and prestressed concrete. BRS makes the generalization that vibration must become unpleasant or painful to the occupants before there is any possibility of damage of the building itself.

Much of the effort in deriving damage criteria has been in the investigation of damage from blasting. Although in this book we are concerned mostly with steady-state vibration, we can derive some guidance from reports of damage caused by the transient vibrations, for if a certain level is the maximum acceptable for transient vibrations of relatively short duration, then the acceptable level for continuous vibration should be lower.

Those engaged in the use of explosives for constructional work in built-up areas know from experience the severity of explosion (mass and type of charge, distance) that can be used in particular circumstances without causing damage. However, since normally the vibration is well below damage level and vibration measurements are not made, experience of this kind does not yield quantitative criteria.

Information of more value is acquired when tests can be made on buildings that are intended for demolition. The severity of the vibration is progressively increased by increasing the severity of the explosion. The vibration is measured at selected positions and the damage noted. From an analysis of the results of such tests available up to 1961, the United States Bureau of Mines (Duvall and Fogelson, 1961) gave the following levels in terms of the peak particle velocity (ppv) of the ground vibration at the building: major damage 7.6 in./s (193 mm/s), minor damage 5.4 in./s (137 mm/s). Hence, they suggested 2.0 in./s (51 mm/s ppv; 36 mm/s rms) as a reasonable value separating the safe zone from the damage zone.

Great Britain BRS (1970) refers to investigations that indicate that the threshold of damage from ground vibration caused by blasting is represented by a vibration velocity of about 75 mm/s at the foundation wall nearest the blast. More recent work, cited by Gutowski et al. (1978), supports the view that peak ground velocity represents the best simple indicator of potential damage to buildings. On the basis of full scale measurements, usually for blast induced vibration of residential type structures, it appears that the "threshold" of building damage, such as plaster cracking, is in the range 5–51 mm/s ppv. The upper limit agrees with the limit of the "safe" zone recommended by Duvall and Fogelson (1961).

Recognizing the complexity of the factors involved and the paucity of quantitative data, the International Organization for Standardization has issued a draft proposal ISO DP 4866 (International Organization for Standardization, 1975) offering provisional guidance on the measurement and evaluation of vibration in buildings. The document proposes values of vibration velocity that should not be exceeded if damage is to be avoided.

For whole-building vibration, limits are given for vibration from blasting and steady-state vibration in the frequency range 1–100 Hz. For vibration from blasting the velocity is determined at the building foundation nearest the blast, as the resultant of the vertical and two horizontal components derived in the manner detailed in the document. Steady-state vibration is measured as the peak velocity of vertical vibration of floors at the position of maximum amplitude, and of horizontal vibration of the whole building at the upper floors.

The limits given are 3–5 mm/s for "threshold damage" and 5–30 mm/s for "minor damage." Threshold damage is defined as visible cracking in non-structural members such as partitions, facings, and plaster walls. Minor damage is defined as visible cracking in structural members such as masonry walls, beams, columns, and slabs, with no reduction in load-carrying capacity. These two categories are indicated on the left side of Fig. 4.4 at their equivalent rms levels.

For the steady-state vibration of a floor in a building, in the frequency range 10–100 Hz, the document indicates that damage is not possible if the peak velocity is below 2.5 mm/s and that there is low probability of damage if the velocity is below 10 mm/s.

The wide range given by various investigators for the damage threshold level is understandable, bearing in mind that the value purports to be applicable to buildings in general regardless of the type or age of the building, the kind of construction, the quality of materials and workmanship, the thickness and span of floors and walls, the duration of the vibration, or its characteristics other than velocity. Present work on damage criteria (Gutowski et al., 1978) recognizes that the damage threshold has a statistical distribution and seeks to assign a probability that damage of a certain kind will result from vibration of a certain velocity.

SENSITIVE EQUIPMENT

The term *sensitive equipment* in this book denotes vibration-sensitive equipment, which includes any equipment whose performance is deteriorated by ambient vibration received through its support. The term is applicable to a wide range of instruments and equipment used in research, teaching, testing, and manufacture.

We now discuss the ways in which various kinds of equipment are disturbed, the need for site vibration criteria, and the applications and limitations of existing criteria.

The nature of the disturbance of sensitive equipment varies with the kind of equipment and its function. With instruments having direct-reading optical indication, such as analytical balances and galvanometers, ambient vibration disturbs the optical system and consequently the reference line or scale appears blurred. With electron microscopes, photographic enlargers, copying cameras, and other instruments that display optical images for visual examination and photography, ambient vibration deteriorates the sharpness of the image and hence the quality of detail that can be observed and photographed.

The accuracy of measurement with optical interferometers is determined by the sharpness and stability of the pattern of interferometric fringe images that is observed. Research scientists using interferometric techniques sometimes find to their annoyance and frustration that their apparatus functions more efficiently as a vibration detector than in the manner intended. Ambient vibration causes relative vibration between the optical surfaces that produce the fringes, with the result that the fringes also vibrate. A relative displacement of the surfaces of only about a thousandth of a millimeter is sufficient to make the fringe displacement greater than the fringe spacing, rendering the pattern unreadable. Vibration can be even more troublesome with interferometers in which a pool of mercury is used to provide a horizontal reference surface (e.g., Example 5.4). A mercury surface is so responsive to vibration that in the days before sensitive portable vibrometers were available, and incidentally before the toxic effects of mercury were recognized, the rippling of mercury in a shallow dish was commonly used as a vibration comparator to find the least disturbed site for sensitive equipment.

Many measuring instruments today incorporate sensitive electrical transducers as detectors or sensors. The electrical signal containing information about the physical quantity being measured may be partially or wholly swamped by "hash" or "noise" caused by ambient vibration.

With some equipment used in research and testing, the ill effects of vibration are indirect. For example, in microhardness testing with very light loads, vibration of the indenter relative to the test piece increases the depth of indentation, with the result that the test material appears to be softer than it is. In experimental research requiring the attainment of extremely low temperatures approaching the absolute zero, ambient vibration introduces unwanted mechanical energy into the apparatus, which has the effect of raising the temperature.

Vibration at the site of a precision machine tool such as a fine grinding machine causes vibration of the cutting head relative to the workpiece, thereby deteriorating the quality of the surface finish and introducing dimensional faults. In the finish grinding of large rolls, ambient vibration can produce ripple marks on the roll surface. When the roll is used subsequently in the cold rolling of metal sheet, these imperfections impart a visible ripple pattern to the finished product which then cannot be marketed as first quality. However, not all unwanted vibration in metal cutting is attributable to ambient vibration. The so-called "chatter" in metal cutting is a much more complicated phenomenon that occurs under conditions determined by the dynamics of the machine tool frame, the geometry of the cutting tool, and the cutting conditions.

A few decades ago there was little demand for environmental vibration criteria for sensitive equipment. If and when trouble occurred, an isolating mounting was devised and if necessary modified and adjusted until it was effective. This was and still is a reasonable procedure for a small item but not for large, heavy equipment such as an electron microscope, a map-copying camera, or a roll grinder. The cost of redesign and reconstruction of an unsatisfactory installation, and the cost of lost production while this is being done, may be many times greater than the cost of providing a mounting in the original installation.

Clearly, the need for vibration isolation should be foreseen in planning the installation of sensitive equipment in existing buildings and in the design of buildings in which sensitive equipment will be used. A decision must be made whether a seismic mounting is necessary. The right decision is more likely to be made if criteria of acceptable vibration are available.

A new and vital need for vibration criteria has arisen in relation to the equipment installations in nuclear power stations in areas subjected to earthquakes. Manufacturers of electrical and mechanical equipment are being required to establish the "seismic qualification" of equipment for use in such installations, that is, to verify by shock and vibration testing under simulated earthquake conditions that their equipment will not merely survive but continue to operate satisfactorily if subjected to a seismic disturbance of a specified magnitude. In this book we are interested in the response of equipment to much lower levels of vibration: the ambient vibration that occurs in buildings under normal conditions of occupation and usage.

The most direct and realistic indicator of acceptable ambient vibration is the equipment itself. If the equipment is portable and can be operated at the various possible sites, an experienced observer will be able to judge whether the ambient vibration is acceptable or not. For larger equipment quantitative criteria are required. Existing criteria are based on vibration measurements made in the course of investigations of particular installations, by correlating observations of equipment performance and observed site vibration.

There have been some attempts to determine criteria by observing the behavior of the equipment under conditions of controlled vibration; for example, the response of a microhardness tester to impacts applied to the frame (Kennedy and Marrotte, 1969) and the response of equipment on a suspended floor to footfall-induced vibration (Ungar and White, 1979).

Vibration tables are available which are capable of vibrating test objects such as spacecraft and missiles weighing some tonnes. From vibration and shock testing, a great deal of information has been acquired about the performance of various kinds of equipment used in space, air, sea, and land vehicles, when subjected to simulated environmental conditions. Although the same testing facilities could be used to determine vibration criteria for a wide range of instruments and equipment in stationary applications, there seems to be no evidence of their use for this purpose.

One would expect that the manufacturers of sensitive equipment would recommend environmental vibration limits within which their equipment would operate satisfactorily. Only a few do so, notably the manufacturers of certain makes of microbalance, electron microscope, and precision machine tool. Some examples of vibration tolerances for particular equipment are given by Steffens (1969) and Ferahian and Ward (1970).

The ISA Criteria

For sensitive equipment in general, the Instrument Society of America (ISA) has made recommendations (Isaacs et al., 1964) for various conditions including vibration along with temperature, humidity, dust, noise, and other environments.

The recommendations are for the laboratories of aerospace and other industries, universities, armed forces, and instrument manufacturers, which maintain reference standards that have been calibrated by the national laboratory and are then used to check the working standards. These laboratories are referred to as Echelon II. The ISA recommendation for vibration at the instrument base is that the acceleration be less than $0.001\ g$ ($0.01\ \text{m/s}^2$) and the displacement amplitude less than 1 microinch ($0.025\ \mu\text{m}$) in the frequency range below 200 Hz. These limits are shown at their corresponding rms values in the lower part of Fig. 4.4.

The environmental requirements are more demanding for the national standards laboratory (Echelon I) which maintains the standards of measurement for the country, and less demanding for the production-line test departments (Echelon III) where the working standards are used to check the gauges and other instruments used to control the quality of the manufactured products.

Vibration criteria are discussed by Ferahian and Ward (1970) who give examples of acceptable and unacceptable vibration collected over a number of years in the course of investigations of vibration problems with many kinds of sensitive equipment. In the light of this experience, which relates mostly to Echelon II laboratories, they support the ISA recommendations but suggest that the limit be defined with reference to the junction frequency (approximately 100 Hz) between the acceleration and the displacement limits. That is, maximum acceleration $0.001\ g$ above 100 Hz and maximum displacement 1 microinch below 100 Hz, observed at the instrument base. For intermittent vibrations, for example, those resulting from footsteps, they recommend a maximum acceleration of $0.01\ g$ ($0.1\ \text{m/s}^2$).

Vibration measurements were made by Dorien-Brown (1969, unpublished) on a number of basement and suspended floors at the CSIRO National Measurement Laboratory (NML) building at Chippendale, N.S.W., Australia, in 1969, to obtain data to guide the specification of requirements for the new building then being designed for the present site at Bradfield Park, Lindfield, N.S.W. Measurements were made during normal working hours of the vertical vibration of the floor resulting from sources external to the room concerned, including air-conditioning plant, workshop machinery and processes, and electric generators. The vibration was analyzed to determine the amplitude and frequency of each major component. When these were plotted on a nomograph, all results for basement rooms were within the ISA recommended limits (Fig. 4.4).

These results give general support to the ISA recommendations for Echelon II even though NML is an Echelon I laboratory. In basement rooms where the ambient vibration was within the ISA limits for Echelon II, equipment of the kinds used for typically Echelon II measurements operated satisfactorily without seismic mountings, but equipment such as the interferometers used for the Echelon I function of determining the length standard required seismic mountings. On the suspended floors the vibration was generally within limits 0.01 g ($0.1\ \text{m/s}^2$) and 10 microinches ($0.25\ \mu\text{m}$), an order greater than that of the basement floors, and with a person walking on the floor the vibration was several orders greater. The application of criteria to suspended floors is discussed later.

Importance of Frequency in Vibration Criteria

In the introductory paragraphs of this section on Sensitive Equipment, examples were given of ways in which site vibration causes vibration of component parts of sensitive equipment with the result that the performance of the equipment is deteriorated. The magnitude of the vibration of any particular part of the equipment depends on the natural frequency of vibration of that part. If the site vibration frequency happens to be equal or very close to the natural frequency, the response may be several times greater than it would be if these two frequencies were well separated. Because of this, criteria in terms of a limited amplitude of site vibration applicable over a broad frequency range can be misleading, as the following example shows.

Example 4.2.   A surface grinding machine is on a floor that is vibrating vertically as shown schematically in Fig. 4.5a. What is the maximum acceptable displacement amplitude of the floor if the resulting vibration of the wheelhead relative to the workpiece must not exceed 10 μm (0.01 mm) displacement amplitude?

With the machine in operation the vibration of the wheelhead relative to the workpiece is not the result of only site vibration. It is influenced also by the rotation and contact pressure of the grinding wheel, and by the vibration caused by the drives and accessories of the machine itself. For the purposes of this example we can assume that the machine is idle and the wheel not in contact with the workpiece.

We assume also that vibration tests have shown that the frequency response of the wheelhead relative to the workpiece is as shown in Fig. 4.5b. The

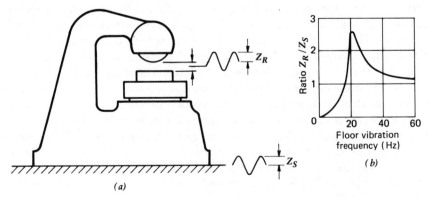

Figure 4.5.   Grinding machine wheelhead response to site vibration (Example 4.2).

response is given in terms of the ratio Z_R/Z_S, where Z_R is the vertical displacement amplitude of the wheelhead relative to the workpiece and Z_S is the vertical displacement amplitude of the floor. Responses of this kind in which the amplitude is magnified in a certain narrow frequency range because of resonance are discussed in detail in Chapter 6.

Referring to Fig. 4.5b, when the floor vibration frequency is in the vicinity of the resonance frequency of 20 Hz, the wheelhead amplitude is greater than that of the floor by a factor of 2.5; therefore, the maximum permissible floor amplitude is 4 μm if Z_R is not to exceed 10 μm. When the floor vibration frequency is well below resonance, say 5 Hz, the wheelhead amplitude is only about one-tenth that of the floor; therefore, a floor amplitude of about 100 μm is permissible. When the floor vibration frequency is well above the resonance frequency, say 60 Hz, the wheelhead amplitude is about the same as that of the floor; hence, the permissible floor vibration amplitude is about 10 μm.

This example shows that the permissible floor vibration amplitude may vary from 4 to 100 μm in the range 5–60 Hz, depending on the value of the floor vibration frequency in relation to the wheelhead resonance frequency.

Because of the possibility of resonance of the components of sensitive equipment, site vibration limits based on observations of a small number of installations of equipment whose resonance frequencies are not known may be misleading. The limits may be too liberal if resonance did not occur during the observations, or too restrictive if resonance did occur and limits were then allocated over a frequency range much wider than that in which the resonance could occur.

A more realistic value of acceptable site vibration could be specified for a particular type of equipment by determining the resonance frequency in any mode of response that may have a significant influence of its performance, and reducing the limit of acceptable site amplitude only in a narrow frequency band containing the resonance frequency. For this more specific kind of criterion to be of practical value, the frequency spectrum of the site vibration must be known.

Sensitive Equipment on Suspended Floors

As mentioned in Chapter 3 in relation to the siting of equipment, an unsupported area of a suspended floor is not a favorable site for sensitive equipment. Nevertheless, laboratory and other vibration-sensitive activities are often located on suspended floors. It is instructive to consider the vibration of a suspended floor caused by a person walking on it, because usually this is the most serious source of vibration.

To estimate the amplitude of floor vibration caused by walking, the most elementary approach is to regard the footfall as a force P (Fig. 4.6a) representing the weight of a walker, applied suddenly to the floor. The resulting vibration

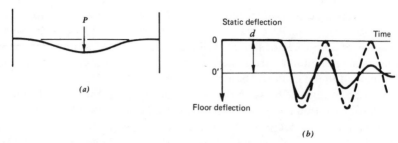

Figure 4.6.   Transient response of suspended floor.

is a transient as shown by the full line in Fig. 4.6*b*. After the transient has decayed, the floor has a static deflection $d = P/k$, where k is the vertical stiffness of the floor.

For the purpose of estimating the maximum amplitude, which will occur during the first cycle, we can consider the vibration to be sinusoidal as shown by the broken line. This vibration takes place about a rest position O′, which is lower than O by the amount of the static deflection. The amplitude of this vibration is equal to the static deflection d (see Fig. 4.6*b*), and the frequency is the natural frequency of the floor. Because the vibration amplitude is linked in this way to the floor stiffness, if we specify an amplitude limit for a suspended floor under footfall loading we are, in effect, specifying a minimum stiffness of the floor.

The specification of the ISA limits for a suspended floor under this footfall loading is inadvisable for the following reason. The vibration frequency concerned is the natural frequency of the floor which, for a suspended floor, is normally well below 100 Hz, in which range the displacement amplitude limit $Z_m = 0.025$ μm governs. Since the displacement amplitude is numerically equal to the static deflection of the floor under the force P, the maximum permissible value of the static deflection is $d_m = Z_m = 0.025$ μm. This condition is satisfied only if the stiffness k is not less than P/Z_m, which calls for an exceptionally stiff floor. For example, calculations made in relation to the design of the new National Measurement Laboratory at Lindfield, N.S.W., showed that a floor of module span 24 ft (7.3 m) would need to be about 1 m thick to satisfy the ISA limit of 0.025 μm with a footfall load of 200 lbf (90.7 kgf).

The "footfall" represented by Fig. 4.6*b*), more aptly called a "heel drop," is an idealization of the force that would occur if a person standing on the toes dropped suddenly onto the heels. The force applied in walking varies with the walker's weight and speed of walking. Ungar and White (1979) have proposed a more realistic form of footstep pulse, idealized from available experimental data for ground loading by footstep. They give data for defining the pulse for a given weight of walker and walking speed, and for calculating the maximum dynamic deflection of the floor resulting from that pulse, given the stiffness and resonance frequency of the floor.

HUMAN SENSITIVITY TO VIBRATION

Most of the research and experiment in this field over the past 50 years has sought to determine the level of vibration that causes discomfort of passengers in vehicles, that interferes with the task performance of drivers or pilots, and that may cause injury of the driver of a tractor, earth-moving, or other industrial machine. More recently, in addition to these studies of the effects of whole-body vibration, there is growing interest in the effects of hand-arm vibration on the operators of hand-held power tools such as pneumatic drills, grinders, rivetters, and chain saws. Investigators are trying to understand the structure and properties of the human body regarded as a mechanical as well as a biological system, and to establish criteria of permissible vibration and means of protection from excessive vibration.

Another topic that is receiving increased attention is "occupant-induced" vibration, which occurs when a person walking on a floor disturbs others on that floor (Allen and Swallow, 1975; McCormick and Mason, 1974; Murray 1979). This is important in the design of buildings for offices, hospitals, libraries, and other purposes where the requirement of large open areas calls for the use of long-span floors of modern lightweight construction.

The perceptibility of vibration depends on the amplitude, frequency, wave-form, duration and intermittency of the vibration, its direction in relation to the body, and the way the body contacts the vibrating surface. Because of differences in temperament as well as in biology, individuals in any group, subjected for test purposes to the same vibration, differ markedly in their assessment of the vibration. Therefore, numerical criteria should be regarded as being no more precise than statistical assessments of orders of magnitude, which may serve as a rough guide for engineering purposes.

For a broad survey of the effects of mechanical shock and vibration on the human body regarded as a mechanical as well as a biological system, tolerance criteria and the protection of humans subjected to various conditions of shock and vibration, reference may be made to von Gierke and Goldman (1976).

The ISO Criteria

In spite of the well-recognized difficulty of establishing a numerical definition of a normal or average sensitivity of humans to vibration, and after nearly 15 years of discussion and development, an international guide ISO 2631 (International Organization for Standardization, 1974c) has been published with the agreement of the majority of ISO member nations. The object of this document is to facilitate the evaluation and comparison of data gained from continuing research, and to give provisional guidance on levels of vibration acceptable to humans under certain defined conditions.

The criteria are applicable to vibration transmitted from solid surfaces to the human body in the frequency range 1–80 Hz. The criteria apply to periodic, random, and transient vibration having its energy within the 1–80 Hz frequency band. The physical factors that determine "vibration exposure" are the intensity or magnitude of the vibration, which is expressed as the rms acceleration, the frequency, the direction in which the vibration is applied, and the duration of exposure to the vibration.

Recommended limits are given for three criteria. The lowest level of vibration intensity is set when the criterion of acceptable vibration is the preservation of comfort. A somewhat higher level represents the limit that will not interfere with working efficiency. The highest level is the limit that should not be exceeded without special justification and precautions, even if no task is to be performed by the individual, to ensure the preservation of health and safety. Limits for these three criteria are given for longitudinal (toe to head, or buttocks to head) and for transverse vibration, for periods of exposure from 1 min to 24 h. In the present context we are interested in criteria for the preservation of comfort. The limit recommended in ISO 2631 for preservation of comfort for 24-h exposure is shown at the left side in Fig. 4.4.

The reduced comfort boundaries are presumed to apply to vibration in transport and near industrial machinery. ISO 2631 notes that in evaluating disturbances due to building vibration in private homes, residential, and office situations, the acceptable level may be below the lowest limit on the ISO criteria, in fact not much above the threshold of perception, especially at night. The threshold of perception in the most sensitive range (1–8 Hz) is considered to be an acceleration level of about 0.01 m/s^2 which is shown on the lower left in Fig. 4.4.

REFERENCES

Allen, D. L., and John C. Swallow (1975). Annoying Floor Vibrations—Diagnosis and Therapy, *Sound Vibration*, **9**(3), 12–17.

Axelsson, Sven-Åke (1968). Analysis of Vibrations in Power Saws, *Royal College of Forestry, Stockholm, Studia Forestalia Suecica*, No. 59.

Den Hartog, J. P. (1956). *Mechanical Vibrations*, 4th ed., McGraw-Hill, New York.

Duvall, W. I., and D. E. Fogelson (1961). Review of Criteria for Estimating Damage to Residences from Blasting Vibrations, *U.S. Bureau of Mines Reports of Investigations*, RI 5968.

Eshleman, Ronald. L. (1976). Vibration Standards, Ch. 19 of *Shock and Vibration Handbook*, 2nd ed., Cyril. M. Harris and Charles. E. Crede, Eds., McGraw-Hill, New York.

Ferahian, R. H., and H. S. Ward (1970). Vibration Environment in Laboratory Buildings, *Canada Natl. Res. Council, Div. Bldg. Res.*, Tech. Paper No. 329.

Great Britain Building Research Station (1970). *Vibrations in Buildings, Part 1*, BRS Digest No. 117.

Gutowski, T. G., L. E. Wittig, and C. L. Dym (1978). Some Aspects of the Ground Vibration Problem, *Noise Control Eng.*, **10**(3), 94–100.

International Organization for Standardization (1973). Balance Quality of Rotating Rigid Bodies, ISO 1940–1973.

International Organization for Standardization (1974a). Mechanical Vibration of Machines with Operating Speeds from 10 to 200 rev/s—Basis for Specifying Evaluation Standards, ISO 2372–1974.

International Organization for Standardization (1974b). Mechanical Vibration of Certain Rotating Electrical Machinery with Shaft Heights between 80 and 400 mm—Measurement and Evaluation of the Vibration Severity, ISO 2373–1974.

International Organization for Standardization (1974c). Guide for the Evaluation of Human Exposure to Whole-Body Vibration, ISO 2631–1974.

International Organization for Standardization (1975). Evaluation and Measurement of Vibration in Buildings, Draft Proposal DP 4866–1975.

International Organization for Standardization (1977). The Measurement and Evaluation of the Vibration Severity of Large Rotating Machines in Situ Operating Speeds from 10 to 200 rev/s, ISO 3945–1977.

Isaacs, A. D., F. G. Klock, Jr., C. D. Koop, J. D. Mitchell, W. F. Snyder, and J. A. Winchell (1964). Recommended Environments for Standards Laboratories, *ISA Trans.* **3**(4), 366–377.

Kennedy, R. G., and N. W. Marrotte (1969). The Effect of Vibration on Microhardness Testing, *Mater. Res. Stand.*, **9**(11), 18–23.

McCormick, M. M., and D. Mason (1974). Office Floor Vibrations Design Criteria and Tests, *Proc. Noise Shock Vib. Conf.*, Monash University, Clayton, Victoria, Australia, pp.198–207.

Murray, Thomas M. (1979). Acceptability Criterion for Occupant-Induced Floor Vibrations, *Sound Vibration*, **13**(11), 24–30.

Muster, Douglas, and Douglas. G. Stadelbauer (1976). Balancing of Rotating Machinery, Ch. 39 of *Shock and Vibration Handbook*, 2nd ed., Cyril M. Harris and Charles. E. Crede, Eds., McGraw-Hill, New York.

Muster, D., and B. Flores (1969). Balancing Criteria and their Relationship to Current American Practice, *ASME Paper* No. 69—Vibr-60.

Rathbone, T. C. (1939). Vibration Tolerance, *Power Plant Eng.*, **43**, 721-24.

Steffens, R. J. (1966). Some Aspects of Structural Vibration, *Proceedings of the Symposium on Vibrations in Civil Engineering*, B. O. Skipp, Ed., Butterworths, London, pp. 1–30; (see also *Gt. Brit. Bldg. Res. Stn.*, *Current Papers Eng.*, Ser. 37, 1966).

Steffens, R. J. (1969). The Problem of Vibrations in Laboratories, *Proceedings of the Conference on Design of Physics Buildings*, University of Lancaster, pp. 25–30, RIBA Publications, London. (See also *Gt. Brit. Bldg. Res. Stn.*, *Current Paper* 14/70.)

Ungar, Eric E., and Robert. W. White (1979). Footfall-Induced Vibrations of Floors Supporting Sensitive Equipment, *Sound Vibration*, **13**(10), 10–13.

von Giercke, Henning E., and David E. Goldman (1976). Effects of Shock and Vibration on Man, Ch. 44 of *Shock and Vibration Handbook*, 2nd ed., Cyril M. Harris and Charles E. Crede, Eds., McGraw-Hill, New York.

5

Seismic Mountings

The support consisted of layers of the following materials: ⅛ in. sheet lead, 4 in. pitch pine, 2 in. nonpareil cork, a No. 24 galvanised iron sheet, 2 in. nonpareil cork, ⅞ in. soft pine board, 4 in. soft pine beams, 1 in. piano-felt pads resting on concrete. . . .

ALEC B. EASON (1923, p. 58)

To this interesting collection of materials one could imagine the witches in Shakespeare's Macbeth wishing to add "eye of newt and toe of frog, wool of bat and tongue of dog. . . . " The quotation is part of a description of a seismic mounting devised some 60 years ago for an air-circulating fan. Fortunately, it is no longer necessary to try various kinds and arrangements of materials in the hope of arriving at some combination that happens to achieve an acceptable degree of isolation. Today we use "off-the-shelf" isolators, which can be selected from ranges of well-established types that have been developed commercially to meet all normal requirements.

A seismic mounting is formed by interposing resilient material between the equipment that is to be isolated and its support. If the equipment is massive and stiff enough it may be placed directly on the resilient material, but usually the equipment is fixed on an "inertia" block, which rests on the resilient material. The inertia block provides additional mass and preserves the alignment of the equipment. The equipment and inertia block together make up the seismic mass.

The literature discloses that many kinds of material have been tried, individually and in combinations, including metal springs, rubber, cork, felt, airbags, squash balls, tennis balls, basketballs, glass fiber, knitted wire mesh, lead, asbestos, soil, sand, gravel, and even a pile of newspapers. Reports have appeared on the use of "popcorn" and chicken feathers for isolating the contents of packages from vibration and impact.

The resilient material may be in the form of a number of separate, individual, or unit isolators such as helical springs, rubber isolators, and air springs. It may be a mat, pad, or other "area" isolator such as rubber, glass fiber, or cork, or a bulk material such as soil or sand. In the discussion to follow, the term *resilient material* includes all three forms. The terms *unit isolator* and *area isolator* are used when necessary to identify these forms; otherwise, we refer to them simply as isolators.

Until about 30 years ago, in general practice there was considerable dependence on "trial and error" because, even if the basic theory were understood, little was known about the dynamic properties of the resilient materials, and instrumentation was not readily available to determine the characteristics of the vibration that was to be isolated. If the resilient material and its arrangement initially tried proved unsuccessful, changes were made until an acceptable degree of isolation was achieved. Details of the final arrangement were of little value to others, except perhaps as an indication of materials and arrangements that they too might try.

Today, dependence on trial and error is not so readily justified. Many kinds of isolator have been developed and marketed for use in seismic mountings. We now have a deeper understanding of the theory, and much more experimental data on the dynamic characteristics of isolators, and instrumentation is commercially available for determining the characteristics of the vibration.

In this chapter, which is concerned with the hardware of seismic mountings, we describe the general arrangements of mountings commonly used, and then discuss resilient materials in general and unit isolators in particular. The design of a seismic mounting involves practical as well as technical considerations: we include a "design guide," which indicates some important practical matters that must be considered. Finally, we discuss the basis of the performance testing of mountings, and the practical difficulties involved.

GENERAL ARRANGEMENT

In designing a seismic mounting it is necessary to decide what resilient material to use and where to place it in relation to the seismic mass. The obvious and normal place is under the mass, an arrangement that is referred to as a "base mounting." The mounting may consist of a massive block formed directly on the ground, or resting on unit or area isolators standing on a structural support. Alternatives to the base mounting are the "center-of-gravity" (c.g.) mounting in which isolators are located around the sides of the seismic mass, nominally in the plane containing the c.g., and the "pendulum" mounting in which the seismic mass is suspended so that its c.g. is below the plane containing the points of attachment of rods or springs that support it.

Base Mounting

Before unit and area isolators came into general use the common practice was to isolate equipment on an independent block, that is, a block not in contact with the surrounding building structure (Fig. 5.1).

An independent block is formed directly on or in the soil, sand, or other site material, which for brevity we shall refer to as the "ground." The installation is, of necessity, located at ground or basement level. An independent block may be used to isolate an engine or other vibration source, or a vibration-sensitive equipment or process.

The vibratory behavior of a massive block resting on the ground is complicated by the fact that the ground functions as both the resilient material and the support. Where does the resilient material end and the support begin? A simplified analysis assumes that some part or "bulb" of the ground adjacent to the block can be regarded as additional seismic mass, and that the material outside this bulb provides the resilience. More advanced theory regards the ground as a material having distributed elastic and damping properties, and deals with the propagation of elastic waves, the dynamic properties of soil, clay, sand, other site materials and combinations of materials, and the way these properties change under the influence of vibration. All of this is outside the scope of the present book. The reader interested in vibration isolation by means of bases

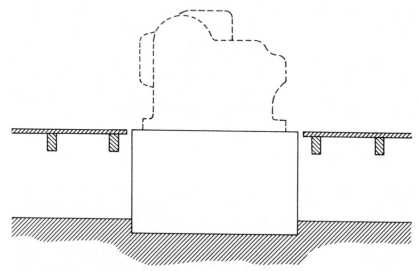

Figure 5.1.   Independent block.

and foundations resting directly on the ground is referred to Barkan (1962), Richart et al. (1970), and British Standards Institution (1974).

The mounting most commonly used today in buildings and industry is the base mounting on unit isolators. The installation may be above the floor, or in a concrete box or pit (Fig. 5.2), which accommodates part of the seismic mass below the floor level so that the equipment remains at a convenient working height.

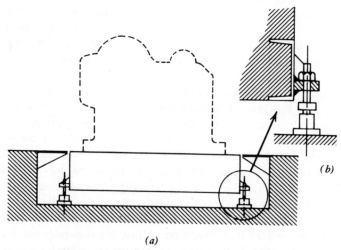

(b)

(a)

Figure 5.2.   Base mounting on rubber isolators.

Example 5.1. A base-type mounting designed to isolate a large roll-grinding machine from vibration generated by cold-rolling mills in a sheet metal manufacturing plant is sketched in Fig. 5.2a. The 40-tonne machine, which can accommodate a 10-tonne roll, is installed on a 67-tonne inertia block. The total seismic mass of 117 tonnes is carried on 54 rubber-in-shear isolators in two rows as illustrated, with a static deflection of 5 mm. The load is applied to the isolators through screws (Fig. 5.2b) so that load sharing among the isolators can be adjusted, and any isolator removed or replaced if necessary.

Center-of-Gravity Mounting

For theoretical reasons that are discussed in Chapter 7, the isolators are sometimes located beside the seismic mass in such a way that when the c.g. of the seismic mass is displaced horizontally the restoring force at each isolator acts in the horizontal plane through the c.g. This may be achieved by using either cross-beams or a T-shaped inertia block as in the following examples.

Example 5.2. In preparation for the installation of a jig boring machine at an engineering works, the vibration was measured at the site of an existing machine that was to be replaced. The vibration resulting from the operation of a plate-shearing press in the vicinity was excessive. Some modification of the foundation of the press made no improvement, so the company, in con-

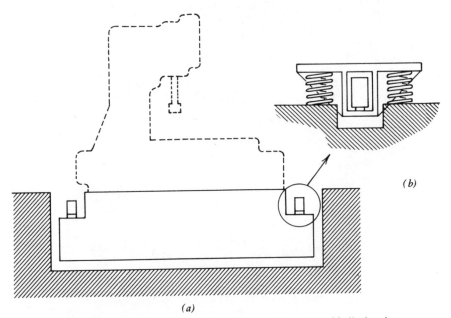

(b)

(a)

Figure 5.3. Center-of-gravity mounting using cross beams and helical springs.

sultation with the author (Macinante, 1955), designed a seismic mounting for the new machine.

The 20-tonne machine was installed on a 30-tonne inertia block (Fig. 5.3) and the whole supported on 16 helical springs in four groups of four arranged symmetrically in plan about the vertical axis through the c.g. of the seismic mass. The load was transferred to the springs through housings at the ends of crossbeams (Fig. 5.3b). The springs, which were made of 38-mm diameter steel, in coils 229-mm mean diameter, were designed for a static deflection of 54 mm, and located in the manner defined above for a c.g. mounting.

Example 5.3.   This example relates to an installation devised for the calibration of accelerometers in terms of the fundamental standards of length and frequency. Basically a calibration involves the measurement of the electrical

Figure 5.4.   Center-of-gravity mounting using T-shaped block on air springs.

output of the accelerometer while it is subjected to sinusoidal vibration of known displacement amplitude and frequency. As the range of calibration that is required in practice involves displacement amplitudes of only a small fraction of a millimeter, special methods—in this example optical interferometry—must be used to determine the amplitude. Unfortunately, a method that is sensitive enough to measure the small calibrating vibration that is imparted to the accelerometer is sensitive also to unintentional and unwanted vibration of elements of the interferometer system; therefore, special attention must be given to the isolation of the calibrator from site vibration.

The interferometric calibrator, developed by Goldberg (1971), is shown in Fig. 5.4 on a seismic mounting designed by Dorien-Brown (1971). The mounting consists of a 1-tonne seismic mass on four air springs. The mass is T-shaped so that the air springs are positioned to satisfy the requirements of a c.g. mounting. With this installation on a basement floor where the vertical vibration was about 0.1 μm at 12 Hz; the vibration of the seismic mass was less than 0.002 μm.

Calibration by interferometry is costly and time consuming. The interferometric calibrator is used to calibrate only high quality *standard* accelerometers, which are then used to calibrate other accelerometers (*reference* accelerometers) by comparison: the standard and the reference accelerometer are subjected to the same vibration and their outputs compared. The reference accelerometer in turn is used in a similar way to check the calibration of "working" accelerometers.

Pendulum Mounting

A type of mounting for small instruments such as galvanometers, which was commonly used during the earlier decades of this century but is rarely used today because it is too cumbersome, consisted of a platform suspended as a pendulum. The platform was attached to wires or springs hanging from the ceiling or from a tall tripod or a wall bracket. Perhaps the best known of these was the Julius suspension in which the platform was provided with a means of adjusting the c.g. of the suspended mass and of damping the free vibration of the system following manipulation or adjustment of the instrument.

A form of pendulum mounting that is sometimes used for large items of equipment is shown schematically in Fig 5.5. The equipment is fixed on an inertia block which is suspended on tension rods. The upper ends of the rods rest on helical compression springs and the lower ends are located in spherical seatings or the equivalent at the ends of through-beams in the inertia block. The pendulum mounting is suitable for large, low-speed engines having unbalanced, horizontally reciprocating parts, because of its low natural frequencies of horizontal vibration.

The literature of about 50 years ago describes some applications of an inverted pendulum for isolating horizontal vibration. For small bench-top in-

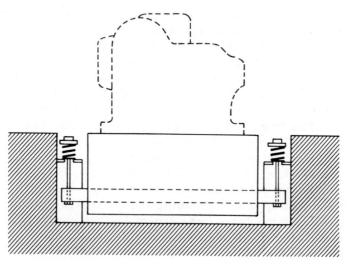

Figure 5.5.   Pendulum mounting on helical springs.

struments such as galvanometers, the instrument table was attached to the tops
of three thin vertical rods so that the system had a very low horizontal natural
frequency. This principle has been applied recently in the following case.

Example 5.4.   A liquid electrometer has been constructed to provide a
precise voltage standard (Clothier et al., 1980). A high voltage is applied
between the surface of a mercury pool and a horizontal electrode located a
few millimeters above it. Electrostatic attraction causes the mercury surface
to rise, and the absolute value of the applied voltage is calculated from the
elevation of the surface and other mechanical properties of the system. The
datum for the elevation measurement is provided by the surfaces of two additional
mercury pools adjacent and connected to the pool to which the voltage is
applied. The accuracy attainable depends to a large extent on the measurement,
by interferometry, of the vertical distance between each mercury surface and
the electrode above it. Therefore it is important to isolate the mercury from
ambient vibration to prevent the formation of ripples. Clothier et al. (1980,
p. 839) mention a number of measures that contribute to a reduction of the
effects of liquid surface vibration: this example refers only to the seismic
mounting designed to minimize disturbance from horizontal vibration.

The three mercury pools are located in a seismic mass of about 1500 kg,
which is supported on three vertical steel rods each 8.7 mm in diameter, 330
mm free length. The natural period of the seismic system may be adjusted by
adding mass; under operating conditions it is of the order of 3 seconds. With
this system a measuring precision of better than 0.3 nm (0.001 interference
fringe spacing) has been attained. With the seismic system rendered inoperative

by clamping the seismic mass, interference observations are impossible because of disturbance of the mercury surface caused by ambient vibration.

ISOLATORS

The most important unit isolators are helical springs, rubber isolators and air springs, and units combining metal spring, rubber, and pneumatic elements. Area isolators currently used include ribbed or embossed rubber carpet, pads or mats of glass fiber, cork, or felt, and composite and layered combinations of such material. Descriptions of the considerable range of isolators commercially available and discussion of their properties can be found in American National Standards Institute (1972), Baker (1975), Crede (1976), and International Organization for Standardization (1972), and in the technical publications of the manufacturers. The earlier part of the discussion to follow refers to isolators in general; then the properties of helical springs, rubber isolators and air springs are considered in more detail.

The properties of importance in the design and choice of isolators are the load-carrying capacity and deflection, the stiffness in the axial and lateral directions, the damping, and the resistance to hostile environments involving exposure to high and low temperature, oil, water, solvents, and nuclear radiation.

The hardware of the mounting and types of isolator are basically the same for source and receiver mountings. The design objective for both, as explained in the next chapter, is to provide suitable values for the natural frequencies and damping of the mounting. The difference between source and receiver mountings is in the considerations, discussed in Chapters 8 and 9, respectively, that determine the desirable values for the natural frequencies and damping.

For a given seismic mass and geometric layout of isolators, the natural frequencies of the mounting are determined by the stiffness of the isolators. The damping requirements are satisfied by selecting a type of isolator that has adequate damping, or by using separate dampers such as those mentioned below under the heading of Helical Springs, which have negligible inherent damping. In the discussion to follow about particular types of unit isolator we give special attention to these two properties, the stiffness and the damping.

A word of caution is necessary about the significance of the term "stiffness" as applied to vibration isolators having nonlinear load/deflection characteristics. In the course of installation of a mounting each isolator receives a share of the weight of the seismic mass and deflects a certain amount, say 10 mm. When the mounting is in service the vibratory displacement of the loaded isolator below and above the mean position is very much smaller, say 0.1 mm. The stiffness that determines the performance of the isolator is not that associated with the relatively large static deflection from the unloaded condition, but that associated with the small oscillatory deflection of the loaded isolator in service.

Furthermore, the stiffness concerned is not the static stiffness in this range of small deflection but the dynamic stiffness, which is the restoring force per unit displacement that the isolator exerts on the seismic mass under the particular conditions of frequency and amplitude of the vibration. For some isolators this dynamic stiffness is significantly different from the static stiffness value that is found by dividing the working load on the isolator by the deflection of the isolator under that load. We discuss this in more detail below under the heading of Rubber Isolators.

The dynamic stiffness can be found only by making tests under dynamic conditions in testing machines designed for the purpose. Data on the properties of isolators under dynamic conditions should be obtainable from the manufacturers of the isolators.

Helical Springs

Helical tension springs were in use as vibration isolators at least a century ago, for example, in the pendulum mountings mentioned earlier for isolating small instruments. Compression springs have been used to isolate machinery for at least 50 years. Today helical spring isolators are almost invariably used in compression. The spring is required not only to support a certain static load without the stress exceeding that allowable for the spring material, but also to have suitable values of axial and lateral stiffness to give the required natural frequencies and to ensure the stability of the mounting.

Within the normal working load range the axial deflection of a helical spring is proportional to the load, that is, the load/deflection relationship is linear and the stiffness is constant. For practical design purposes the stiffness based on a load/deflection calculation or test over a relatively large range of static deflection from the unloaded condition can be taken as the dynamic stiffness appropriate to small oscillatory deflection of the loaded spring in service.

Helical springs are in such general use in engineering that there is no need to reproduce the design formulas or graphs that are used to design a spring to have a specified deflection under a given load. Design data for helical springs intended for use in seismic mountings, and examples of spring isolator units, can be found in Baker (1975), Church (1976), Crede (1976), Spring Research Association (1974), and in technical publications of the major spring manufacturers.

In order to calculate the natural frequencies of a seismic mounting we need to know the axial stiffness and the ratio of the lateral to the axial stiffness. Analysis of the lateral behavior of helical springs is complicated. Wahl (1944, p. 179; see also Wahl, 1963, p. 70) has presented an analysis based on the assumption that as lateral deflection occurs the ends of the spring remain parallel to one another. Crede (1951, p. 245) has simplified Wahl's result for application to a helical spring made of steel of circular cross section and carrying zero axial load. By combining this expression with that for the axial stiffness Crede derives an expression for the ratio of lateral to axial stiffness. In a seismic mounting the springs, of course, carry axial load: Crede gives a formula for a correction factor to account for the influence of axial load on the lateral/axial stiffness ratio.

In many applications it is necessary to design helical compression springs to have low values of axial and lateral stiffness in order to achieve the required low values of the natural frequencies in modes involving horizontal as well as vertical vibration. In seeking to minimize the natural frequency in a horizontal natural mode, we must bear in mind that as the natural frequency approaches zero, the system approaches instability. The conditions of spring design to be satisfied to ensure stability are outlined by Crede (1951, pp. 247–249), and Wahl (1963, pp. 68–70).

Helical springs can be designed to carry a load of a small fraction of a kilogram or more than a tonne. The springs shown in Fig. 5.6 were used to isolate the camera referred to in Example 3.6. The 7-tonne camera was carried on three groups of four springs made of 16-mm square-section steel in coils 117-mm mean diameter. The largest helical spring noted by the author is reported (Blaschke, 1964) to be made of material of diameter 76 mm, mean

Figure 5.6. Helical springs used to isolate a large camera.

coil diameter almost 0.5 m, and free height 1.2 m; its static deflection is 25 mm under a load of 1.2 tonne. These springs were designed to support an entire building in an underground cavern, for use as a command center, to isolate it from shock and vibration in the event of a nuclear explosion.

Helical spring mountings can be designed to have a vertical natural frequency as low as about 2 Hz for which the static deflection is about 60 mm. To attain significantly lower natural frequency the required static deflection of the spring would be excessive. For example, a compression spring for a 1-Hz vertical natural frequency would deflect some 250 mm under the working static load.

The inherent damping in a helical spring isolator is negligible, which is a disadvantage in applications where there is the possibility of unwanted resonance. Any necessary damping must be provided by separate means, such as friction pads or fluid dampers. In fluid dampers a vane or plunger fixed to the seismic mass is immersed with large clearance in a highly viscous fluid in a pot fixed to the base or support. Dampers of the kind used as shock absorbers in vehicles are not used in stationary mountings because usually displacements of one end of the damper relative to the other would be too small to overcome threshold static friction.

Where noise reduction as well as vibration isolation is required in a helical spring mounting, separate noise attenuating materials are inserted at the spring seating. Helical spring isolator units designed for the higher loadings usually contain a nest of springs in a pedestal or housing that may also contain frictional damping and noise attenuating elements. For convenience of installation a helical spring may be preloaded within its housing so that most of the static deflection is taken up before the working load is applied.

In mountings for sensitive instruments, helical springs have the disadvantage that floor vibration can be transmitted through the springs by the vibrations of the coils relative to one another. Wrapping the coils with adhesive tape has been found to reduce the transmission of coil vibrations, but not to an acceptable level for a very sensitive instrument, for which air springs are now used.

Rubber Isolators

Rubber has been used as a vibration isolating material for at least a century. The major developments in rubber isolators followed the discovery that metal could be chemically bonded to rubber, for this made possible the design of unit isolators in which the rubber could be loaded in shear, or combined compression and shear, while bonded to the metal parts necessary for holding the isolator in the required position.

Rubber isolators are commercially available in a very wide variety of shapes and sizes (e.g., see Crede, 1976). Perhaps the most common type is the rubber-in-shear. An example is given in Fig. 5.7, which is a cross section of the type of isolator used in the roll grinder mounting referred to in Example 5.1.

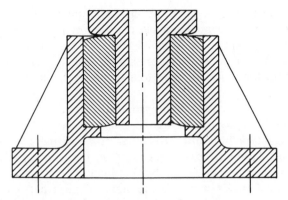

Figure 5.7. Rubber-in-shear isolator.

The design of a rubber isolator is a matter for the specialist, involving the choice of material, natural rubber or a particular composition of synthetic rubber, and the determination of the shape and size of the rubber to achieve the desired load rating and stiffness characteristics.

In practice, the designer of a seismic mounting on rubber isolators makes use of manufacturer's data for the properties of the isolators. These data must be interpreted with care to ensure that the stiffness value used in the design calculations is the true or effective value appropriate to the conditions of actual use, bearing in mind that the static load/deflection relationship is usually non-linear and that the dynamic stiffness is usually greater than the static stiffness. An example is given later (Example 7.2) of the error that can result from neglect of these factors in the calculation of the vertical natural frequency.

The available data may be inadequate, giving only the deflection at the working load, based on the manufacturer's static load/deflection tests. Or worse, these meager data may be misleading if accompanied by a graph or other presentation of a linear relationship between static deflection and natural frequency with no reminder or warning that the linear relationship is valid only for isolators (e.g., helical springs) having a linear load/deflection relationship and no significant difference between the static and dynamic stiffness. The following is an example of the trouble that can result from underestimation of the stiffness of isolators.

Example 5.5. A gear grinding machine sited about 75 m from a forging hammer was installed on a 38-tonne inertia block, and the total seismic mass of 54 tonnes mounted on 32 rubber-in-shear isolators, with a static deflection of 5 mm. The vertical natural frequency corresponding to this static deflection, for an installation on linear isolators, is 7 Hz (see Fig. 7.6). The designer of the installation made some allowance for the effects of nonlinearity and dynamic stiffness, and expected the vertical natural frequency to be about 10 Hz.

The actual value was found to be about 17 Hz, which unfortunately happened to be about the same as the predominant frequency of ground vibration transients resulting from operation of the forging hammer. Vibration measurements made simultaneously on machine and floor showed that the vertical vibratory displacement of the block was three to four times that of the floor.

The effects of nonlinearity and dynamic stiffness deserve some further discussion. Figure 5.8 shows the load/deflection relationship of an isolator having the "hardening" type of nonlinearity, that is, the stiffness increases with the deflection. The static stiffness is not PM/OM, the working load (w) divided by the static deflection (d) caused by that load. The relevant static stiffness (k_s) is given by PM/NM which is the slope of the small part of the load/deflection curve, in the vicinity of the point P, corresponding to the vibratory

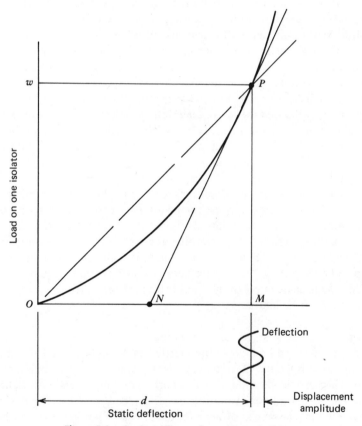

Figure 5.8. Static stiffness of a nonlinear isolator.

displacement of the loaded isolator in service. Obviously this static stiffness may be appreciably greater than the quantity w/d.

The dynamic stiffness (k_d) of the isolator when vibrating may be appreciably higher than k_s. Therefore, the designer of the mounting needs to know for a particular isolator how k_d relates to k_s under the conditions of use. The ratio k_d/k_s is commonly in the range 1 to 2, depending on the strain amplitude in the rubber, the vibration frequency, and the composition of the rubber. Detailed information on the behavior of rubber isolators is given by Allen et al. (1967), Frye (1976), Gobel (1974), Hobaica (1979), and Lindley (1974).

In comparison with helical springs, rubber isolators have the advantage that they have enough inherent damping for most applications. Damping is desirable in a source mounting to limit the response under resonance conditions, and in a sensitive equipment mounting to suppress the free vibration. The internal damping depends on the rubber compound used, and the vibratory strain and frequency. Typical values indicate that the damping ratio may be as high as 0.2, although in practice it is commonly less than 0.1 (Crede, 1976, Chap. 32, pp. 23–24). Rubber isolators also attenuate noise, which metal springs readily transmit unless they are used in conjunction with noise attenuating materials. However, the vertical natural frequency attainable with rubber isolators is not much below about 10 Hz, even when two isolators are used in series at each isolator position, which gives helical springs the advantage when a vertical natural frequency below about 10 Hz is required. It should be noted too, in relation to the isolation of high frequency vibration (some hundreds of hertz), that the efficiency of rubber isolators is diminished at the particular frequencies at which standing waves occur in the rubber.

Bridge Bearings

Today natural and synthetic rubber isolators are commercially available in compositions that are suitable for wide ranges of environmental conditions, and in load ratings suitable for isolating a small instrument, a large machine, a bridge, or a whole building. The largest rubber isolators, which were developed as bridge bearings, are in the form of a block of natural rubber containing a number of steel plates arranged horizontally and bonded to the rubber to form a multiple sandwich as shown schematically in Fig. 5.9. The outer plates with their relatively thin protective coating of rubber are the bearing surfaces between which the unit is loaded in compression. The internal steel plates have the effect of increasing the compressive but not the shear stiffness. An isolator of this type mentioned by Waller (1969, p. 74) carries a load of 220 tonnes with a static deflection of 10 mm. The British Standards Institution (1975) have published a document bringing together, in a form suitable for everyday use by structural engineers, practical information on the design, operation, and performance of isolators of this type.

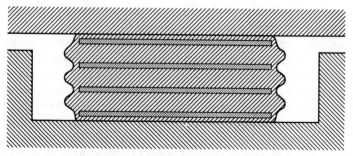

Figure 5.9.   Bridge-bearing type of isolator.

Air Springs

An air spring is simply a container of compressed air which supports a load. Air springs have been in use and development as vibration isolators over the past 50 years. In an early application, devised to isolate a small instrument such as a galvanometer, the air spring was an inner tube of a motor vehicle tire. The partly inflated tube was placed on a horizontal support and a platform carrying the instrument was placed on the tube. In another early application, the floor of a broadcasting studio was supported on an array of rubber bags connected to a compressed air supply. The system was designed to isolate the floor from the vibration of other parts of the building and to permit the room acoustics to be adjusted by controlling the pressure in the groups of air bags supporting separate sections of the floor.

Air springs of the kinds used today to isolate machinery and sensitive equipment in buildings have evolved since about the 1960s (Gieck, 1962) from the types developed over the preceding two or three decades for use in vehicle suspensions. The air springs used in vehicle suspensions are of a bellows type, which does not have appreciable lateral stiffness; the springs form part of a system that does not depend on the air spring for lateral stability. In stationary installations, however, the isolators alone support the seismic mass; therefore, they must have adequate lateral stiffness to ensure stability.

Today, commercially available air springs are of three basic types, shown schematically in Fig. 5.10. A wealth of technical information, design data, and examples of practical applications is available from the leading manufacturers.

Bellows Type

Figure 5.10*a* is a cross section of a two-convolution air spring. The envelope is made of nylon-reinforced rubber, sealed to hold pressure typically up to about 700 kPa, and there are metal plates at top and bottom for locating and loading the air spring and through which the air is supplied. A small air spring is suitable for a design load of about 50 kg, and a large one, having an outside diameter of say 0.5 m, about 15 tonnes. Figure 5.11 shows one of the air springs referred to in the following description of a seismic table designed by Dorien-Brown (1971) in collaboration with K. M. O'Toole.

Example 5.6. A 320-kg seismic table 1.8 × 1.5 m for optical holography is supported on four air springs of the two-convolution bellows type, about 200 mm diameter. A high quality seismic table was essential for this work because the only room available had a suspended concrete floor which vibrated vertically with 0.5 μm p-p displacement due to activities elsewhere in the building, and much higher if an observer moved about the room. On this table highly satisfactory holograms were made which required exposure times up to 2.5 min, during which time the relative displacement of critical parts of the optical system must not exceed about 0.1 μm.

In the design of a mounting on air springs, as on isolators of any other kind, it is necessary to know both the axial stiffness and the ratio of lateral to axial stiffness of the isolator (Chapter 7). The axial stiffness can be found from the manufacturer's data graph relating load and height for a given pressure. From a given initial condition (height, load, pressure) the change in height for an assumed small increase in load is read from the graph and the axial stiffness calculated as the rate of change of load with height. The lateral stiffness of the single-convolution type is about one-half the axial stiffness. The lateral stiffness of two and three convolution types is less than that of the single convolution. The lateral stiffness data are normally supplied by the air spring manufacturer.

Air Spring with Height Control

The vertical load on an air spring is equal to the product of the "gauge" pressure of the contained air and the effective area of the air spring. During a vertical vibration, when the mass moves down a small distance from the rest position the height of the air spring decreases by a small amount. This reduces the volume, increases the pressure, and provides an upward force tending to restore the mass to its rest position. The relationship between restoring force and deflection varies with the shape and material of the air spring envelope, and can be estimated by using the theory of air compression, or determined by making tests on the air springs.

An important conclusion can be drawn from a simplified theory based on the assumption that the effective area is constant; that is, the air spring is assumed to behave as a piston/cylinder element. By considering a small downward displacement of the piston and using the laws of gas compression to derive the increase in pressure associated with the reduction in volume, it turns out that the stiffness of the air spring, and hence the natural frequency of a given mass supported on the spring, depends only on the height of the air spring. A consequence of this is that if the load is changed and the pressure adjusted to keep the height the same for all loadings, the vertical natural frequency is the same for all values of the load. The theoretical basis of vibration isolation using air springs with servo control is discussed by Cavanaugh (1976).

Automatic height control, which is obviously advantageous in vehicle suspensions, is valuable also in mountings for stationary equipment; for example, machine tools and weighing equipment which must be kept horizontal when the load and/or its distribution is changed. Air isolators have been developed (Kunica, 1965) which incorporate a height sensing device to control the air supply so that when the load is increased air is admitted to the air spring, and when the load is reduced air is released, thereby maintaining the spring at a constant height. The basic form of this type of isolator, shown schematically in Fig. 5.10b, consists of a metallic body or cylinder, which acts as an air chamber, and can support the load when it is not supported by the air pressure. There is a flexible seal or diaphragm between the cylinder and the "piston" that carries the payload. The diaphragm operates nominally at zero deflection

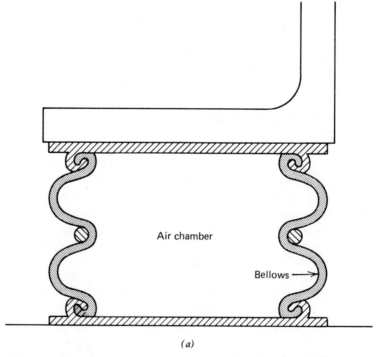

Air chamber

Bellows →

(a)

Figure 5.10.   Schematic diagrams of air springs: (a) Two-convolution bellows type.

because the load is maintained at a constant height by a valve that admits air when the piston is below and releases air when the piston is above the mean position.

With air springs a natural frequency of a few hertz can be attained with "off the shelf" isolators and without the inconveniently large static deflection that is involved when helical springs are used to achieve a natural frequency of 3 Hz or lower. Air springs offer the further advantage that by increasing the volume of air by means of a supplementary or "surge" tank, the natural frequency can be made lower than that attainable if the volume is limited to

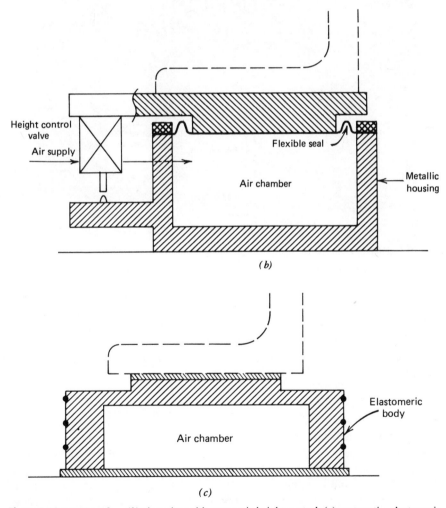

Figure 5.10 (*continued*). (*b*) air spring with automatic height control; (*c*) pneumatic–elastomeric isolator.

Figure 5.11.   Air spring isolator for optical holographic table.

that of the air spring alone. In this way a vertical natural frequency of about 1 Hz is practicable. By contrast, a mounting on helical springs having a vertical natural frequency of 1 Hz is unthinkable because the springs would have a static deflection of about 250 mm, and correspondingly excessive height and diameter.

Pneumatic–Elastomeric Type

The theoretical and laboratory investigation of the prototype isolator of this type is described by Schubert (1974). The main features, shown schematically in Fig 5.10c, are an elastomeric thick-walled cylindrical body and a top shaped as a diaphragm coupling the body to the top plate. Unlike the thin wall of the bellows type, which serves only to contain the air, the thick wall of the pneumatic–elastomeric type can act as a temporary mounting pad during installation of the equipment, or at any other time when the isolator is not inflated. The thick wall also acts as a snubber if a vertical shock causes a downward displacement greater than that of the compliant top. The transverse stiffness is about equal to the axial stiffness, providing good transverse isolation and stability. Steel rings around the outside prevent bulging, and improve the deflection and stability characteristics. Isolators of this type are now commercially available in a wide range of load ratings.

Damping of Air Springs

There is some inherent damping in the bellows and the pneumatic–elastomeric types, due to straining of the flexible air chamber. Damping of an air spring can be increased by making use of the surge tank mentioned earlier as a means of attaining a lower natural frequency. This additional damping is generated by restricting the flow of air between the air chamber and the surge tank. By making this restriction variable, the amount of damping can be adjusted to the optimum for the particular application. In the example in Fig 5.11 (Example 5.6) the air spring is connected to a surge tank, which serves also as a pedestal support for the air spring. There is an adjustable capillary between the air spring and the surge tank. If for some reason (e.g., height limitation) the surge tank is located remotely from the air spring, a restriction may be provided in the interconnecting air line. An adjustable capillary damper for this purpose has been designed by J. J. Mouttou (1975, unpublished).

To sum up, clearly the air spring has many attractive features. It offers natural frequencies of 1 Hz and lower which are impracticable with any other form of passive isolator; it is readily adaptable for use with a height sensing valve to maintain the height, level, and natural frequency of the mounting when the load or its distribution changes. On the debit side, air springs require regular inspection to ensure that the air pressure is maintained. The mounting is designed so that if for any reason the air spring becomes deflated the seismic mass will descend only a small distance and come to rest on a structural or other support, for example, the screw jack in Fig. 5.11.

Another feature that may be objectionable when air springs are used to isolate sensitive equipment is that a change in ambient pressure (as might result from the sudden opening of a door) causes a transient vertical displacement of the equipment resulting from the sudden change in pressure differential across the flexible envelope.

DESIGN GUIDE

We assume that, after consideration of the various matters discussed in Chapter 3 in relation to action at the source, action at the receiver, and vibration insurance, it is decided to provide a seismic mounting, and that the site chosen is a favorable one in accordance with the considerations discussed in Chapter 3.

This design guide is not a "do-it-yourself" recipe or check list for the design, but an outline of what has to be done. The decision of "who does what" depends on the size, complexity, and importance of the proposed installation. It may be advisable to retain the services of a consulting engineer who specializes in vibration, or of one of the major suppliers of vibration isolators, who are experienced in the design and application of their products, and offer services ranging from technical advice to the design, supply, and installation of complete isolating systems.

The guide refers first to the considerations that determine the type of mounting (i.e., choice of isolator, general arrangement) and then to certain important practical considerations, specifically those relating to access to the isolators, flexible connections, stiffness of the inertia block, and stiffness of support pillars.

There are other design requirements of a practical nature including access for cleaning and drainage of the pits of below-floor mountings, "housekeeping" pads or plinths for above-floor mountings, and corrosion protection of metallic parts. Useful guidance on these and other details that require attention in the design and specification of seismic mountings is given by U.S. General Services Administration (1971), and ASHRAE (1980). Reference may be made also to ANSI (1972) and ISO (1982), which aim to facilitate communication between user and manufacturer of isolators.

Type of Mounting

The type of isolator and general arrangement of the mounting are determined by the required natural frequencies of the mounting and other considerations mentioned in the earlier parts of this chapter. The desirable natural frequencies are found by using the procedures given under the heading Practical Application of the Design Data in Chapter 8 for a source mounting, and in Chapter 9 for a sensitive-equipment mounting.

In following these procedures, it may be necessary to use vibration measurements to determine the characteristics of the vibration generated by a machine, and the natural frequencies of a suspended floor. For a sensitive-equipment mounting, measurements may be necessary to determine the site vibration and the natural response characteristics of the equipment.

The method of achieving, in the hardware, the natural frequency that is theoretically desirable, is given in Chapter 8 under the heading Select Isolators and Arrange Layout.

Access to Isolators

The isolators must be accessible for inspection, and for adjustment or replacement if this becomes necessary. Examples of the trouble that can result from failure to provide access are given in Example 3.8, where it was not possible to inspect the isolators, and Example 3.4, where the isolators were readily inspected but could not be aligned or replaced.

The design should be such that either the load can be raised off the isolators by using lifting tackle or jacks, or the isolators may be unloaded individually. An example of the latter is seen in Fig. 5.2, where the load is applied through screws, which can be withdrawn to permit alignment or removal of any isolator, and adjusted so that the isolators are equally loaded. Ample space is allowed between the inertia block and the sides of the pit for access to the isolators.

The isolators supporting a "floating floor" can be made accessible from above the floor. The slab is formed leaving holes for the isolators. Subsequently the slab is raised onto the isolators by using screws in cover plates fixed over the holes.

Most types of isolator have a metal pedestal or housing that can support the load if the isolating medium fails. When using other types (e.g., air bellows) provision should be made for the load to be transferred to a suitable support if the isolator fails (deflates). The alternative support may be an abutment as in Fig. 5.9 or a screw jack as in Fig. 5.11. Also with the bellows type, side stops should be provided so that the lateral deflection cannot exceed the permissible amount.

Flexible Connections

The design of the seismic mounting is based on the assumption that the seismic mass is coupled to the support only through the isolators. In practice, almost invariably there are other connections, through pipes, hoses, and cables, carrying air, gas, water, and electricity. These connections must be flexible, so that the motion of the seismic mass is unconstrained, and the transmission of vibration and noise between equipment and supporting structure is minimized.

The importance of flexible connections to sensitive equipment is well illustrated in a case study (Crawford, 1969), which shows that these connections can transmit more vibration to the equipment than do the isolators.

Example 5.7. A 23-tonne vacuum chamber about 6 m long, and 1.8 m diameter overall, was supported on a pneumatic isolation system designed for a vertical natural frequency 0.75 Hz or less, and horizontal natural frequencies 2.5 Hz or less. The designers recognized the importance of flexible connections and made the provisions considered necessary. When the system was turned on, the vibration of the vacuum chamber was too great for the intended application. After lengthy testing and analysis of every possible vibration source and transmission path (there were some 30 connections) modifications were made to certain couplings. On retesting the system, its performance exceeded the requirements.

Severe vibration of a seismically mounted machine can cause trouble with piping connections as well as malfunction of the machine itself. This is particularly so if the machine has angular as well as translational vibration, because of the magnification of the amplitude with distance from the axis of rotation. The vibration can cause loosening of pipe joints, and even cracking of pipes. Fortunately, we seldom experience the more serious consequences referred to by Barkan (1962 p. xi).

Example 5.8. ". . . due to an unsatisfactory design, the foundations of several piston gas compressors installed in the same building underwent considerable vibrations. The compressors, being rigidly connected to the foundations, also underwent vibrations of large amplitude which were transmitted to the pipes leading from them. Vibrations frequently caused damage to and even rupture of these pipes near the vibrating compressors. This caused the seepage of explosive gas into the compressor house; as a result the house blew up."

Stiffness of Inertia Block

The design basis given in this book for the calculation of natural frequencies (Chapter 7), and for deciding what the natural frequencies should be (Chapters 8 and 9), assumes that the seismic mass is a rigid body. In practice this is a reasonable assumption for a small inertia block, and for a large block that is not long in comparison with its cross-sectional dimensions. An inertia block may be made stiffer by increasing its depth. An example of such an inertia block for a large grinding machine is given by Allaway (1960).

The need for adequate stiffness of the inertia block is most obvious when a long optical bench is to be installed on a seismic mounting. As the length of the inertia block may be ten times its width or depth, its flexural vibration should be considered. The following example relates to a case study by Crawford (1969), and a paper by Krach (1968), on the design of long reinforced concrete beams supported on soft isolators.

Example 5.9. A reinforced concrete beam 15 m long, 1.2 m wide, and 1.4 m deep was required to support an optical equipment. A long beam supported on soft isolators vibrates in a free–free mode. The natural frequency calculated for a beam of these dimensions in the fundamental free–free mode is 17.8 Hz. If this beam were simply supported, the value would be 7.8 Hz. The fundamental frequency was raised to 20.3 Hz by casting a cylindrical void 760 mm diameter through the full length of the beam. This design feature, besides increasing the stiffness and natural frequency, reduces the seismic mass. Consequently the floor loading, the load rating of the isolators, and the installation cost all were reduced.

Stiffness of Support Pillars

For the installation of small items of sensitive equipment it is sometimes convenient to use a pillar bolted and grouted to the floor to support either the equipment or the isolators of an above-floor mounting. In the author's experience such a pillar, although its proportions and method of fixing may appear to be satisfactory, invariably magnifies the horizontal vibration.

On several occasions the author has observed the vibration amplitude at the top of an instrument support pillar to be three to five times that at the floor near the base of the pillar. Because instrument-support pillars are short and squat, the magnification results from rocking on the base rather than flexure as a cantilever.

Example 5.10. On one occasion the observed frequency at the top of a pillar was about 50 Hz, and the waveform almost sinusoidal, of varying amplitude. After some time was spent in a fruitless search for a 50 Hz source, perhaps a motor or a vibrator operating in the vicinity, further observations were made. From a record of a manually induced free vibration of the pillar, it was found that the natural frequency was close to 50 Hz. The pillar was acting as a mechanical filter for the random vibration of the floor and magnifying frequencies in the vicinity of its natural frequency.

PERFORMANCE TESTING

The logical sequel to the design and installation of a seismic mounting is its performance testing, to compare its actual behavior with that intended by the designer. The author is not aware of any recognized code of practice for the performance testing of seismic mountings. Commonly, an installation, of which the mounting forms part, is accepted if it functions satisfactorily and without objectionable vibration.

Thus, for a source mounting, the criterion is that the vibration produced by a machine be acceptable to the operator of the machine and to those working or living in the vicinity. The following example (Baxa and Ebisch, 1978) is of interest for its application of this criterion in relation to an extremely severe source of vibration, and for the subtle timing of the acceptance test.

Example 5.11. A company engaged in the crushing and shredding of scrap metal for recycling planned to install a 4000 hp hammer mill-type of automobile shredder. On learning that another company in this business was facing costly litigation arising from the operation of a similar plant, the company provided a carefully designed seismic mounting for the shredder. The hammer mill and its motor were installed on a 363-tonne inertia block, and the total load of 500 tonnes supported on 16 air springs of the two-convolution bellows type, with a vertical natural frequency of 1 Hz. The system was designed to isolate the very severe unbalance that would result from the loss of two hammers, each 190 kg at 1.2 m radius, normally rotating at 600 rev/min, and the shock resulting from the occasional explosion in the shredder of an automobile gasoline tank.

When opponents of the project protested, at the city council meeting, that if the shredder were allowed to operate it would shake the neighborhood apart, the company representative had the pleasure of informing the meeting that the shredder had been in operation for 2 weeks and nobody had noticed the vibration.

However effective an acceptance test of this kind, which is concerned only with the effects of the vibration transmitted through the mounting, it is not, in a technical sense, a performance test of the mounting, for which it is necessary to compare the transmitted (output) amplitude of force, or displacement, with the corresponding applied (input) amplitude. The experimental evaluation of a mounting on this basis presents some difficulty, which we now discuss.

Source Mountings

The purpose of a source mounting is to attenuate alternating forces, so that the force amplitude transmitted into the supporting structure is appreciably smaller than that generated by the source. The experimental determination of a "force transmissibility" as the ratio of the amplitudes of two alternating forces is not practicable. Although it may be feasible, for the purpose of experimental research or investigation to measure the forces generated by a machine (e.g., with force transducers designed into bearings), and the force transmitted by the mounting (e.g., with dynamic load cells under the isolators), the use of such methods in normal practical applications is unthinkable. This situation appears to have caused little concern because normally it is not the force but the resulting vibration that matters.

The performance of a source mounting can be determined by making only vibration measurements, if the vibration at a specified vibration-sensitive area can be measured with and without the mounting. This is obvious and attractive in principle, but in practice can be done only with installations of trivial size, such that the isolators can be removed and the equipment fixed to the floor for the test. For larger installations this is impracticable not only because of the labor and cost involved but also because of the difficulty of providing temporarily, for the purpose of the test, an effectively "rigid" fixing, in place of the isolators.

To sum up, a source mounting is designed on theory relating to force transmissibility, as discussed in Chapters 6 and 8, but it cannot be assessed experimentally on this basis. In the present state of the art, the performance of the mounting per se is not determined (which, incidentally, is sometimes a bonus for the designer). Instead, the total installation, including the mounting, is assessed on the criterion that, with the source operating under specified conditions (e.g., speed, load), the vibration it causes at specified vibration-sensitive areas must be within specified limits, examples of which are given in Chapter 4.

Mountings for Sensitive Equipment

As with vibration sources, sensitive equipment installations are usually assessed only on overall performance. The installation is accepted if the equipment attains the required standard of performance or output, for example, a particular grade of surface finish produced by a grinding machine, a desired sharpness of the image in an electron microscope, or simply freedom from vibration of the reference line or scale in a measuring instrument.

The function of a seismic mounting is to minimize the "internal" vibrations of the equipment that deteriorate the quality of the output. Therefore, the performance testing of the mounting should involve a comparison of the vibration of the critical parts of the equipment with that of the site.

If the critical elements of the equipment are accessible, or if the consequences of their vibration are measurable (e.g., the depth of imperfections in a ground surface; the width of a blurred line in an optical image), then the performance of the mounting can be expressed as the ratio of the magnitude of the unwanted response to that of the site vibration. This form of transmissibility ratio is defined and discussed later (Chapters 6 and 9). The following are examples of a method used by the author to measure the response of the critical elements of precision machine tools. Today measurements of this kind can be made with commercially available transducers of various kinds.

Examples 5.12.   In tests on a jig borer (Example 5.2) the relative vibration between tool spindle and worktable, with machine idle, resulting from normal and test excitations applied to the site, was measured with an air-gap capacitive transducer and associated displacement meter designed by H. A. Ross (see Macinante, 1955, Fig. 10). One plate of the transducer was supported in the tool holder in the spindle, and separated by a small air-gap from the other plate which was attached to an angle plate on the worktable.

In tests on a roll grinder (Example 5.1) an air-gap capacitive transducer was set up to measure the relative vibration between wheel and roll, with machine idle, in response to site excitations. The plate was attached to the surface of the wheel, separated by a small gap from a similar plate attached to the surface of the roll.

In principle, the performance of the mounting can be determined by measuring the response of a critical element of the equipment, to a given site excitation, with and without the isolators. The author tried this method during the tests on the jig borer referred to in Example 5.2. Hardwood blocks were fixed with steel wedges between the inertia block and the pit, and the motion of the tool spindle relative to the worktable of the machine was measured with a certain test excitation (dropped weight) and compared with the response to the same excitation with the inertia mass "floating" on the isolators. However, vibration measurements made on the wedged block and the adjacent floor, with the test excitation, showed that the displacement of the block was appreciably less than that of the floor. Therefore, the test was abandoned because, if the wedging had been effective, the vibration of the block would have been the same as that of the floor.

In the tests of the gear grinder mounting referred to in Example 5.5, an attempt was made to nullify the isolators by using steel screw jacks to clamp the seismic mass. With 14 jacks tightened around the sides of the block and 17 under the block, the displacement of the block in response to a forging hammer blow was appreciably greater than that of the adjacent floor. Again the test was abandoned because the wedging was inadequate.

While the wedging in both of these cases may have been satisfactory if displacements of the order of one or two millimeters were involved, it was ineffective in these cases where the displacements to be compared were well below 0.1 mm. At such small displacements the blocks or jacks, in effect, merely function as stiffer isolators on which the seismic mass can vibrate.

An interesting recent application of this kind of performance test is described by Kelly and Chitty (1980).

Example 5.13.   In relation to the protection of essential equipment such as pumps, piping, valves, and control devices in power plants in seismically active regions, the feasibility of constructing the entire plant on a base-isolated system was studied by testing an experimental model of a three-story steel frame, 6 m high, and of total weight 20 tonnes, supported on elastomeric bearings on a 6 × 6 m shaking table. The response of elements of the model representing critical elements of the plant was measured with and without the isolators. For the latter condition the isolators were bridged by welded steel straps.

To sum up, in normal practice it is likely that the acceptance testing of sensitive equipment installations will continue to be based on the satisfactory performance of the equipment and not on a specific evaluation of the mounting. This is reasonable from the viewpoint of the owner or user of the equipment, but provides no "feedback" for the designer of the mounting. The experimental determination of the quality of the mounting, expressed as a ratio of quantities representing the unwanted response and the site vibration, is quite practicable with existing vibration measuring techniques and instrumentation. However, the cost of testing of this kind may not be justified in normal practice, for its only purpose would be to verify the design of the mounting.

REFERENCES

Allaway, P. H. (1960). Foundation for Roll Grinding Machines, *Engineer*, **210**, 457–458 [see also *Noise Control*, **7**(1), 30–32, (Jan.-Feb.), 1961].

Allen, P. W., P. B. Lindley, and A. R. Payne (1967). *Use of Rubber in Engineering*, Natural Rubber Producers' Research Association, Maclaren, London.

American National Standards Institute (1972). *American National Standard Guide for Describing the Characteristics of Resilient Mountings*, ANSI S2.8., 1972.

American Society of Heating, Ventilating and Air-Conditioning Engineers (1980). *ASHRAE Handbook and Product Directory 1980 Systems*, Ch. 35, Sound and Vibration Control.

Baker, J. K. (1975). *Vibration Isolation*, Great Britain Design Council, British Standards Institution, and Council of Engineering Institutions, Engineering Design Guide No. 13, Oxford University Press.

Barkan, D. D. (1962). *Dynamics of Bases and Foundations*, McGraw-Hill, New York.

Baxa, Donald E., and Robert Ebisch (1978). Controlling Automobile Shredder Vibration through Pneumatic Isolation, *Noisexpo Conference 6th Proc.*, pp. 24–28.

Blaschke, Theodore O. (1964). Underground Command Center, *Civil Eng.* **34**(5), 36–39.

British Standards Institution (1974). *Foundations for Machinery*, Part I Foundations for Reciprocating Machines, CP 2012: Part I.

British Standards Institution (1975). *Vibration Isolation of Structures by Elastomeric Mountings*, Draft for Development DD 47.

Cavanaugh, Richard D. (1976). Air Suspension and Servo-Controlled Isolation Systems, in *Shock and Vibration Handbook*, 2nd ed., Chap. 33, Cyril M. Harris and Charles E. Crede, Eds., McGraw-Hill, New York.

Church, Austin H. (1976), Mechanical Springs, in *Shock and Vibration* Handbook, 2nd ed., Chap. 34, Cyril M. Harris and Charles E. Crede, Eds., McGraw-Hill, New York.

Clothier, W. K., G. J. Sloggett, and H. Bairnsfather (1980). Precise Reflection Interferometry System for an Absolute Standard of Voltage, *Opt. Eng.*, **19**(6), 834–842.

Crawford, Robert W. (1969). Vibration Isolation of Large Masses and Vacuum Chambers, Optical Telescope Technology Workshop NASA SP-233, pp. 733–749.

Crede, Charles E. (1951). *Vibration and Shock Isolation*, Wiley, New York.

Crede, Charles E. (1976). Application and Design of Isolators, in *Shock and Vibration Handbook*, 2nd ed., Chap. 32, Cyril M. Harris and Charles E. Crede, Eds., McGraw-Hill, New York.

Dorien-Brown, B. (1971). Air Springs for Vibration Isolation, *Harold Armstrong Conference on Production Science in Industry*, Institution of Engineers Australia, Volume of Preprints, pp. 213–225.

Eason, Alec B. (1923). *The Prevention of Vibration and Noise*, Henry Frowde and Hodder & Stoughton, London.

Frye, William A. (1976). Rubber Springs, in *Shock and Vibration Handbook*, 2nd ed., Chap. 35, Cyril M. Harris and Charles E. Crede, Eds., McGraw-Hill, New York.

Gieck, J. E. (1962), Design with Air Springs, *Product Eng.*, **33**(24) 63–74.

Göbel, E. F. (1974). *Rubber Springs Design*, Newnes–Butterworths, London.

Goldberg, J. L. (1971). An Interferometric Method for the Standardization of Oscillatory Displacement, *Metrologia*, **7**(3), 87–103.

Hobaica, E. C. (1979). Behaviour of Elastomeric Materials under Dynamic Loads II, *Shock and Vibration Digest*, **11**(7), 11–18.

International Organization for Standardization (1982). *Vibration and Shock Isolators—Procedure for Specifying Characteristics*, ISO 2017–1982, 2nd ed.

Kelly, James M., and Daniel E. Chitty (1980). Control of Seismic Response of Piping Systems and Components in Power Plants by Base Isolation, *Engng. Struct.*, **2**(3), 187–198.

Krach, F. G. (1968). Reinforced Concrete Beam Resonances, *Shock Vibration Bull.*, **38**(2), 133–138.

Kunica, Serge (1965). Servo-Controlled Pneumatic Isolators —Their Properties and Applications, *ASME Paper* 65—WA/MD—12.

Lindley, P. B. (1974). *Engineering Design with Natural Rubber*, 4th ed., Malaysian Rubber Producers' Research Association, London.

Macinante, J. A. (1955). The Measurement and Isolation of Vibration, *J. Inst. Eng. Aust.*, **27**, 323–337.

Richart, F. E., Jr., J. R. Hall, Jr., and R. D. Woods (1970). *Vibrations of Soils and Foundations*, Prentice-Hall, Englewood Cliffs, New Jersey.

Schubert, Dale W. (1974). Dynamic Characteristics of a Pneumatic-Elastomeric Shock and Vibration Isolator for Heavy Machinery, *Inst. Environ. Sci. Proc.* 20th Annu. Meet., p. 244–251.

Spring Research Association (1974). Helical Springs, Great Britain Design Council, British Standards Institute, and Council of Engineering Institutions, Engineering Design Guide No. 8, Oxford University Press.

U.S. General Services Administration (1971). Public Buildings Service, Guide Specification, Vibration Isolation, PBS 4-1515-71.

Wahl, A. M. (1944). *Mechanical Springs*, Penton, Cleveland.

Wahl, A. M. (1963). *Mechanical Springs*, 2nd ed., McGraw-Hill, New York.

Waller, R. A. (1969). *Building on Springs*, Pergamon, London.

6

Design Model for a Seismic Mounting

It is perfectly obvious, of course, that the harder one looks at any physical process the more complex it appears. The art of the engineer lies (at least partly) in knowing when to stop peering at things and to start "getting on with it."

R. E. D. BISHOP (1979, p. 162)

Some of the earlier seismic mountings were successful, others not. Also not all recent installations are satisfactory, because there is still some dependence on trial and error: an uninformed or careless choice or layout of isolators can result in increased rather than decreased transmission of vibration. The design of a mounting on the basis of formulas and charts applied without a clear understanding of their physical relevance to the installation and of their limitations can lead to trouble.

Formulas used in the design of seismic mountings are expressions describing the response of a "design model" of the installation to a certain excitation. In this chapter we describe the simplest, the one-mass model, and its response to periodic excitation. We indicate ways in which this model does not adequately represent the installation, and show that these inadequacies may be remedied by the use of a two-mass model.

We then look critically at the two-mass model and see that it, too, has limitations which may be removed by further refinement of the model. This leads to a consideration of the role of design models in general. Bearing in mind that as the complexity of the model increases so also does that of the analysis of its dynamic response, we must decide how much refinement of the model is justified. In other words, what is the trade-off between the advantage of having a simple model and the penalty of using a model that does not adequately represent the installation that is being designed?

After considering this question in some detail, we adopt the two-mass model, which, although it has certain limitations, is much more realistic than the one-mass model that is still commonly used as the basis for the design of mountings. Use of the two-mass model enables the designer of the hardware to "get on with it" without the guesswork involved in the use of the one-mass model, and with the expectation that the two-mass model should be adequate for many of the installations met in normal practice.

The discussion to follow examines in some detail the merits of the two-mass model for installations subjected to periodic excitation. Design models for installations subjected to shock, transient, and random excitation are of the same general nature, the number of subsystems and their arrangement being determined by the particular installation. The designer must decide on a suitable design model, analyze its response to the particular excitation, and determine values of the design parameters that satisfy whatever design criterion applies. Some comments on design models for installations subjected to these other forms of excitation are given later under the heading Practical Application of Two-Mass Model.

ONE-MASS MODEL OF A SEISMIC MOUNTING

Designers of mountings some 50 years ago, aware that vibration isolation could be achieved by using resilient material, visualized the mounting as the mass–spring system shown schematically in Fig. 6.1a. The system represented by a diagram of this kind for the purpose of designing a mounting is called a design model of the installation.

The equipment, represented by the "lumped" mass m, is supported on isolators represented by a spring of stiffness denoted by k, which is the vertical stiffness of the resilient material interposed between the mass and the support. The spring is assumed to be linear, that is, its load/deflection relationship is a straight line. The stiffness k is the slope of the load/deflection line expressed in units of force per unit deflection. It is assumed that only vertical motion is possible and as a reminder of this sometimes "guides" are drawn, as in Fig. 6.1, but not subsequently.

Referring now to installations subjected to periodic excitation; when this one-mass system is used as the model for a source mounting, an alternating force acts on the mass as in Fig. 6.1b, and when the model represents a receiver mounting the excitation is a vibration of the support as in Fig. 6.1c. Vibration associated with external or applied excitation as in Fig. 6.1b and 6.1c is called "forced" vibration when it is necessary to distinguish it from "free" vibration that can occur in the absence of excitation.

First considering source isolation, let us assume that a sinusoidal vertical force of amplitude P_S acts on the mass as in Fig. 6.1b. At very low frequencies the force in the spring and hence the force transmitted to the support is equal to the applied force, that is, the amplitude (P_T) of the transmitted force is equal to P_S. As the forcing frequency is increased it will attain a value that coincides with the natural frequency, which is the frequency of the vertical free vibration that would occur if the mass were suddenly released from a displaced position. When this coincidence occurs the system is in resonance and the amplitude of vertical displacement of the mass, and hence the amplitude of the force transmitted through the spring, is much greater than that of the applied force. With further increase in frequency the resonance response subsides and the transmitted force amplitude decreases.

A similar effect occurs when the excitation is an alternating displacement of the support as in Fig. 6.1c. This can be seen in the following simple ex-

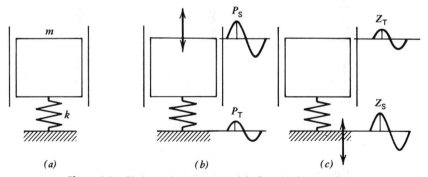

Figure 6.1.   Undamped one-mass model of a seismic mounting.

periment, which is made with a spring in tension as in Fig. 6.2 because it is easier to perform than with a spring in compression.

Take a light spring or a strip of rubber, and at one end attach a mass sufficient to stretch the spring about 50–100 mm. Clamp or hold the other end stationary, pull the mass down and release it, and note that its natural frequency in vertical free vibration is about 2 Hz. Now hold the upper end, as sketched in Fig. 6.2*a*, and move the hand up and down very slowly. The spring will not extend or contract noticeably, and the displacement of the mass will be about the same as that of the hand. Then gradually increase the frequency of vertical motion of the hand until resonance occurs and the mass oscillates vigorously up and down (Fig. 6.2*b*). Then go on increasing the frequency and observe that at frequencies well above the resonance frequency (Fig. 6.2*c*) the vertical displacement of the mass is negligible even though the hand moves rapidly up and down. The spring is now isolating the mass from the vibration of the hand.

These obvious changes in the amplitude of displacement of the mass are accompanied by changes in the phase of the motion that are not obvious. At frequencies below the resonance frequency the mass moves in phase with the hand. In passing through resonance the phase changes so that at frequencies well above the resonance frequency the mass and the hand are in opposite phase, or antiphase. In the design of seismic mountings the difference in phase between excitation and response is of no importance.

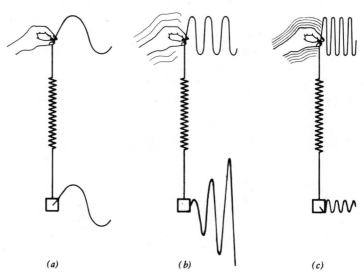

(a)   *(b)*   *(c)*

Figure 6.2.   Simple demonstration of vibration isolation.

Obviously the response of the simple mass–spring system depends very much on the excitation frequency, which is expressed as the frequency ratio (f/f_n), the ratio of the excitation frequency (f) to the natural frequency (f_n) of the system.

With excitation applied to the mass (Fig. 6.1b) the performance of the system as a vibration isolating mounting is expressed as the force transmissibility ratio T_F, which is the ratio (P_T/P_S) of the amplitude of the transmitted force to that of the applied force. With excitation applied to the support (Fig. 6.1c) the performance is expressed as the displacement transmissibility ratio T_D, which is the ratio (Z_T/Z_S) of the displacement amplitude of the mass to that of the support.

It is a remarkable fact that the expression derived for T_F of a source mounting turns out to be identical to that for T_D of a receiver mounting. The proof of this can be found in textbooks (e.g., Den Hartog, 1956, Chap. 2; Crede, 1951, Chap. 2). The expression is given in Eq. (6.1) and shown graphically in Fig. 6.3.

$$T_F = T_D = \frac{1}{1 - (f/f_n)^2} \tag{6.1}$$

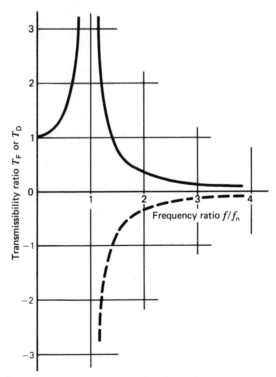

Figure 6.3.   Transmissibility of undamped one-mass system.

It may be noticed that Eq. (6.1) yields negative values for T_F or T_D when f/f_n is greater than unity. This is a consequence of the sign convention adopted in deriving the expression and is consistent with the physical fact that the transmitted and applied forces in Fig. 6.1b are in phase below resonance and antiphase above resonance. Likewise the displacement of the mass in Fig. 6.1c is in phase with that of the support below resonance and antiphase above resonance. In practice only the magnitude or absolute value of the transmissibility ratio is of interest; therefore, the negative sign is ignored and the negative part of the response curve, shown as the broken line in Fig. 6.3, is drawn on the positive side as shown by the full line.

The results of the experiment illustrated in Fig. 6.2 are consistent with Eq. (6.1) and Fig. 6.3. At frequencies well below resonance (Fig. 6.2a), for example, with $f/f_n = 0.1$, $T_D \approx 1$; that is, the amplitude of the mass is about the same as that of the hand. At frequencies well above resonance (Fig. 6.2c), for example, with $f/f_n = 4$, $T_D \approx 0$; that is, the amplitude of the mass is negligibly small compared with that of the hand.

However, there are two ways in which the behavior of the model in Fig. 6.1 is unrealistic. First, it can be seen by substituting $f/f_n = 1$ in Eq. (6.1) that the transmissibility ratio T_F or T_D at resonance is infinite, which in the hardware is unthinkable. Second, a free vibration of the system, once initiated, theoretically goes on forever with undiminished amplitude.

The model behaves in this unrealistic manner because it does not incorporate any element that dissipates energy. In any real system there is some damping or energy dissipation resulting from internal friction in the isolators, and friction at any rubbing surfaces. The term "damping" connotes energy dissipation and should not be used, as it sometimes is, as a vague general synonym for "vibration isolation." Damping limits the resonance response in forced vibration and causes free vibration to decay. For these reasons damping is intentionally designed into mountings.

To overcome this deficiency in the undamped model, theoreticians introduced the concept of a "viscous damper" acting in parallel with the spring as shown schematically in Fig. 6.4. Other arrangements, notably elastically supported damper systems (Ruzicka and Cavanaugh, 1958) have certain advantages; in

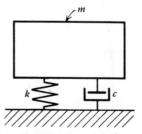

Figure 6.4.   Damped one-mass model of a seismic mounting.

the present context we refer specifically to the conventional arrangement Fig. 6.4. The damper may be thought of as something like the piston/cylinder element used in automobile shock absorbers and door closers, although in practice this type of damper is not used with seismic mountings. In the model the damper is supposed to provide damping equivalent to that of all the energy dissipating mechanisms in the system.

The damper is assumed to resist motion of one end relative to the other with a force proportional to the velocity of the relative motion. The constant of proportionality (c), the damping coefficient, is the force required to move one end of the damper with unit velocity relative to the other end. The original and continuing attraction of this concept of a "velocity" or "viscous" damper is that it is easy to handle mathematically, and yields results of practical value in the applications referred to in this book. Various other damping characteristics have been defined and their influence on the behavior of vibrating systems determined. Ruzicka and Derby (1971) present design data collated from published results of scores of investigations of the response to sinusoidal excitation of the one-mass system having various types of damping (viscous, Coulomb, quadratic, hysteritic, velocity–nth power, combined viscous and Coulomb).

With the inclusion of damping, the behavior of the one-mass system becomes much more realistic. The free vibration now decays instead of going on interminably, and the response at resonance is no longer infinite but is limited by the damping. The derivation of expressions for the force and displacement transmissibility of the damped system can be found in textbooks (e.g., Crede, 1951, Chap. 4; Den Hartog, 1956, Chap. 2), which also discuss the phase relationship between the excitation and the response. As with the undamped system, the expression for the force transmissibility ratio T_F turns out to be the same as that for the displacement transmissibility ratio T_D. The expression need not be reproduced here; it is used only by students when obliged to do so in solving tutorial problems. In practice graphs are used, such as that in Fig. 6.5, which shows transmissibility ratio versus frequency ratio for various conditions of the damping.

The damping is expressed as the damping ratio (c/c_c) where c is the damping coefficient defined earlier, and c_c is the coefficient of critical damping, which is defined as follows. If the system in Fig. 6.4 has damping less than critical the free vibration is oscillatory; if greater than critical, the mass if displaced creeps back to the rest position and does not overshoot. The critical damping coefficient is that for the limiting condition separating oscillatory from nonoscillatory motion.

Because the function of a seismic mounting is to ensure that the amplitude of the transmitted vibratory force or displacement is significantly less than the excitation amplitude, the design aim is to achieve values of the transmissibility ratio T_F or T_D well below unity. Such values are attained only for values of f/f_n well above the value, theoretically shown to be $\sqrt{2}$, where the various curves intersect in Fig. 6.5. In other words, the aim is to have a "soft"

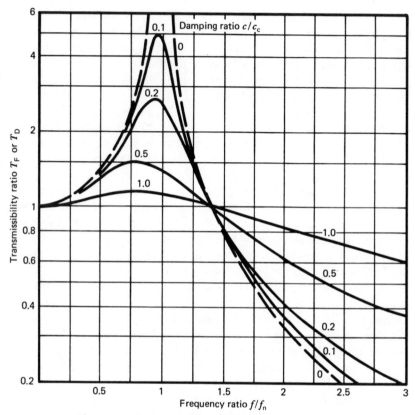

Figure 6.5.   Transmissibility of damped one-mass system.

mounting, that is, a mounting whose natural frequency f_n is much lower than the excitation frequency f.

In practice, only rarely is the excitation purely sinusoidal. Usually it is periodic, and the natural frequency of the mounting is designed to be much lower than the fundamental frequency because it will then be a smaller fraction of the frequency of any harmonic present in the periodic vibration. In other words, the fundamental frequency is taken to be the excitation frequency.

Failure to make f/f_n greater than $\sqrt{2}$ results in a mounting that at best achieves no isolation. This happens if the mounting is much too stiff thereby making f_n much higher than f ($f/f_n \approx 0$). At worst the mounting natural frequency happens to be equal to the resonance frequency ($f/f_n \approx 1$) and the mounting magnifies the vibration to a degree that depends on the damping.

So far in this discussion the design model has been considered to have freedom of motion in the vertical direction. The same model and the conclusions about its response to sinusoidal excitation can be applied to a system having only horizontal freedom of motion when excited by a sinusoidal horizontal excitation, and to a system having only rotational freedom when acted on by a sinusoidal angular excitation.

The behavior of the damped one-mass system illustrated by the curves in Fig. 6.5 is very well known. It may not be too much of an exaggeration to say that the practical design of mountings is based almost entirely on this model, even though it has the following serious limitations.

As a representation of a source mounting, the design model in Fig. 6.4 is illogical because the support is represented as rigid and unlimited. Therefore, it cannot transmit vibration, which must mean that a source isolating mounting is unnecessary.

As a model for a mounting for a sensitive equipment represented by the rigid mass in Fig. 6.4, the model is realistic only if the function of the mounting is to minimize vibration of the equipment as a whole resulting from vibration of the support. However, a mounting for sensitive equipment is usually required to minimize the response of some critical part or element of the equipment relative to some other part, in which case the equipment cannot logically be represented as a rigid mass.

Despite these inadequacies, mountings designed on the basis of the damped one-mass model are not necessarily doomed to failure. The inadequacy of the design model causes no trouble with a source mounting if the support happens to be stiff enough, or with a sensitive equipment mounting if the critical elements of the equipment happen to be stiff enough. How stiff is "stiff enough" will become evident in Chapters 8 and 9.

TWO-MASS MODEL AND ITS LIMITATIONS

Clearly, then, the one-mass model does invite some refinement. The model for a source mounting should be made capable of representing the flexibility of the support, and that for a sensitive equipment capable of representing the important component parts of the equipment in such a way that their responses may be investigated. The least we can do to satisfy these needs is to provide a second mass–spring–damper system in the following ways.

For a source mounting we can replace the rigid support of the one-mass model with a mass–spring–damper system representing the support, for example, the suspended floor in Fig. 6.6. For a sensitive equipment we can represent a critical or responsive element of the equipment, for example, the cutting head of a machine tool, by a secondary mass–spring–damper system, as shown in Fig. 6.7.

The two-mass model formed in this way overcomes the deficiencies mentioned in relation to the one-mass model, but both the one-mass and the two-mass models have other deficiencies. We now look critically at the two-mass model and ask some questions pointing to these deficiencies.

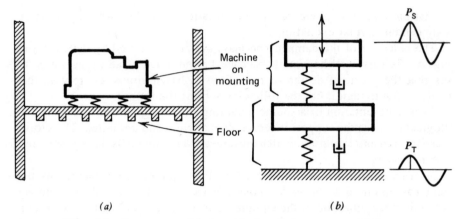

Figure 6.6.   Two-mass model of machine installation on suspended floor.

Machine as Rigid Mass

In the model for a source mounting (Fig. 6.6b) how realistic is the use of a rigid mass to represent the entire machine that is the source of vibration? For example, if the source is a rotating machine of some kind, the excitation derives from mass unbalance of rotating parts. Figure 6.8a shows a typical rotor. The centrifugal force resulting from unbalance is transferred to the base of the machine through the shaft and bearing supports, all of which have properties of mass, elasticity, and damping. Consequently, the alternating force input to the mounting may differ appreciably in amplitude and phase from the centrifugal force resulting from unbalance. Should we represent a shaft/rotor system such as that in Fig. 6.8a as a mass-spring-damper system as in Fig. 6.8b so that its influence on the transfer of the excitation from rotor to mounting can be taken into account?

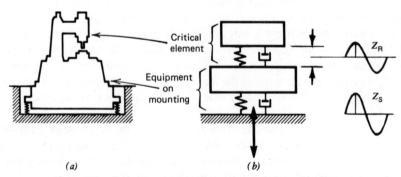

Figure 6.7.   Two-mass model of sensitive-equipment installation.

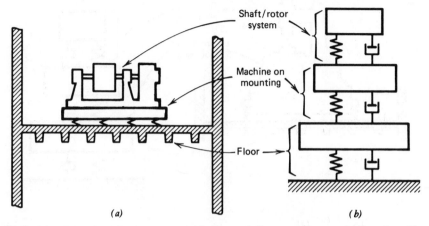

Figure 6.8. Extension of two-mass model in Figure 6.6*b* to represent nonrigidity of machine.

Also the flanges or feet by which the machine is fixed to the inertia block, if not well designed, may need to be treated as spring elements separating the two masses (machine and inertia block) which, in Fig. 6.6*b*, are represented as a single lumped mass. Flexibility of the mounting flanges or feet of a machine may cause deterioration of the quality of isolation, as discussed by Snowdon (1979, p. 37).

Excitation at Support of Sensitive Equipment

For the purpose of designing a mounting using the model in Fig. 6.7*b*, the vibration of the support must be defined to have a certain amplitude and frequency. How is this excitation to be determined? It cannot be taken simply as the vibration measured at the support before installing the equipment because the vibration of the unloaded support caused by a given source will not be the same as that produced by the same source when the support is loaded by the installation. Should we represent the support, for example, the suspended floor in Fig. 6.9*a*, as a mass–spring–damper system as in Fig. 6.9*b* so its influence on the transmission of vibration to the mounting from an external source can be portrayed by the model?

Responses in Directions Other Than Vertical

So far we have considered refinement of the model by increasing the number of mass–spring–damper systems while retaining unidirectional motion of the masses. If the motion were unconstrained each mass may have six degrees of freedom, as discussed in some detail in Chapter 7.

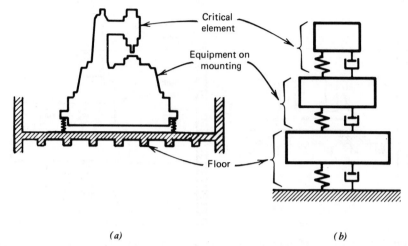

(a)   *(b)*

Figure 6.9. Extension of two-mass model in Figure 6.7*b* to represent nonrigidity of suspended floor.

The unidirectional model of a source installation is limited in its application because actual machinery generates forces having horizontal as well as vertical components, and couples about various axes; consequently , the seismic mass and the support may well have horizontal and angular motions as well as vertical. Similarly, the unidirectional model of a sensitive equipment installation cannot be used to examine responses of the seismic mass or critical elements in directions other than the vertical. By restricting consideration to unidirectional motion are we unduly limiting the usefulness of the model?

Actual installations rarely have the kind of symmetry that is implied in unidirectional design models. The assumption of unidirectional motion seems to imply that the models are applicable only to installations that "behave themselves," their vibrations never deviating from the straight and narrow path. Will the unidirectional model be of value in relation to installations that lack this symmetry?

Nonlinearity of Stiffness and Damping

In the basic model we assume that the isolators have linear properties: stiffness proportional to displacement and damping proportional to velocity. Isolators other than helical metal springs may be markedly nonlinear when tested over their range of working load. Should we consider the isolators in the model to have nonlinear stiffness and damping?

Lumped or Distributed Properties

In the models we have used lumped masses. Even the suspended floor, despite its obviously distributed mass, elasticity, and damping, is represented by a lumped mass–spring–damper system. In actual installations mass is not localized in lumps, and the elasticity and damping are not confined to certain parts. All of these properties are distributed throughout the materials. Why not allow these elements in the models to have distributed properties?

Summing Up

We have noted several ways in which the two-mass unidirectional model appears to be oversimplified. We have raised a number of questions that cast doubt on the usefulness and relevance of the model, and hinted at ways of refining it to make it more realistic. But where do we stop? How far should we go in refining the model? Or, in Bishop's words at the beginning of this chapter, when do we stop peering at things and start "getting on with it?" In order to answer this question we must be clear on the role of the design model.

ROLE OF A DESIGN MODEL

A design or mathematical model is a conceptual, idealized representation of the essential features of a mechanical or other physical system. The model is devised to exclude characteristics or features of the hardware and modes of its response to excitation that are of minor importance. It involves the use of simplifying assumptions which deliberately and drastically reduce the number of "degrees of freedom," that is, the number of coordinates (parameters, variables) necessary to define the configuration of the system at any instant. For example, the system in Fig. 6.1 that can vibrate only vertically has one degree of freedom, but if the mass were unconstrained it would have six degrees of freedom. The two-mass unidirectional system in Figs. 6.6 and 6.7 has two degrees of freedom, one associated with the vertical position of each mass, but if the masses were unconstrained the system would have 12 degrees of freedom. The greater the simplification the less information the model will yield. The model should be capable of representing those features of the system that are important in the context of the particular application of the model.

One might ask why it is necessary to strive for simplification of the model; why not allow it to represent all features and modes of response of the hardware, and subsequently ignore those that are shown by theoretical analysis and computation to be unimportant? A few decades ago the reason for simplification was that the computations involved, even for systems with only two or three degrees of freedom, were too lengthy and tedious for the computational facilities then available. For this reason Crede (1951, p. 148) declined to compute the response to shock excitation of the two-mass system in Fig. 6.7b.

With the advent and general availability of high speed digital computers the "number crunching" now presents no difficulty. The complexity of systems being designed today and the extent of the dynamic analysis undertaken are indicated by Hager (1980) in a review of the history of dynamic analysis since 1947. For example, in the design of the NASA Space Shuttle a comprehensive analysis was essential because the major structural loadings would occur only during actual launch and landing, when the project could not be stopped for redesign. The analysis required the assumption of a very large number of degrees of freedom associated with many possible modes of dynamic response. Hager states that today the capability exists for 20,000 degrees of freedom and 850 dynamic modes, and that in the future the capability will be unlimited. The constraints now are imposed not by computational difficulties but by the time required to get the data into the computer, and the task of digesting and interpreting the enormous quantities of computer output.

Another reason for simplifying the model is that the more complicated the model the more input data it demands about the hardware. For example, when the one-mass model in Fig. 6.4 is extended to the two-mass forms, data are required for the mass, stiffness, and damping of the structure supporting the source mounting (Fig. 6.6), or of the critical elements of the sensitive equipment (Fig. 6.7), in addition to the mass, stiffness, and damping data for the mounting.

The art in devising a design model is to know how far to carry the simplification. It follows that one should not put too much trust in the model until its adequacy and validity have been established by testing the actual system that has been modeled.

An example of the experimental assessment of a design model is described by the author and colleagues who made a vibration study of a large steerable radio telescope (Macinante et al., 1967) with the object of comparing its dynamic behavior with that predicted by the designers. The experimental work disclosed 10 significant modes of vibration. The simplest design model capable of representing these modes was then described (Macinante, 1969) and is available for the design of other structures of this kind.

It is easy, of course, to see in the light of hindsight, having observed the actual behavior of the structure, what the design model should be. In practice normally the model must be decided before the structure exists, and this calls for experience and judgment.

CASE FOR ADOPTION OF TWO-MASS MODEL

We recall that the one-mass model was set aside because it cannot represent the flexibility of the support, which is important when a source mounting is installed on a suspended floor, and it cannot represent the flexibility of the critical elements of sensitive equipment whose responses determine the success or failure of the mounting for the equipment. We have shown that a two-mass model can represent these features but it has other limitations, which have been discussed.

Bearing in mind the assertion that the design model should be the simplest that is capable of representing the important features of the installation, the author has adopted for detailed presentation in this book the two-mass model in the forms shown in Figs. 6.6b and 6.7b.

There can be no doubt that this two-mass unidirectional model is the simplest that could be contemplated for the purpose, but whether or not it adequately represents all the important features of the installation depends on the importance attached to the various features that may require representation. The author's view of the case for adopting the two-mass model is given below under headings corresponding to those used earlier when questions were raised concerning the adequacy of the two-mass model.

Machine as Rigid Mass

The two-mass model in Fig. 6.6b does not represent the nonrigidity of the machine that is the source of the vibration.

Because of nonrigidity, for example, that of the shaft/rotor system (Fig. 6.8a) or the holding-down flanges or feet, the alternating force input to the mounting may differ in amplitude and phase from the vertical component of the dynamic unbalance generated by operation of the machine. The maximum effect would occur at a resonance, when the rotational frequency coincides with a natural frequency of the shaft/rotor system or with a natural frequency of the whole machine on its holding-down bolts.

In the author's experience, serious resonances of these kinds are uncommon. A machine is normally designed so that, although the shaft/rotor system may pass through one or more of its critical (resonance) speeds in coming to its intended continuously-running speed, the latter will be well separated from any critical speed. Likewise, good design practice demands adequate stiffness of the machine-fixing flanges or feet. However, if the machine is of an unusual kind and resonances are expected, the design model may have to be refined so that the influence of the resonances on the transmissibility, and hence on the design of the mounting, may be investigated.

In the method presented in Chapter 8 involving the use of the two-mass model (Fig. 6.6b), the transmissibility ratio (T_F) is defined as the ratio P_T/P_S of the transmitted to the applied force amplitude. The applied force P_S is nominally the vertical component of the unbalanced force generated by the machine. In practice this force cannot be determined directly by measurement and is rarely available as data from the manufacturer of the machine. All that the designer of the mounting can do is to estimate P_S from a knowledge of the machine dynamics and the dynamic balance quality.

This design value of P_S is applied to the lumped seismic mass of machine plus inertia block (Fig. 6.6a). Nonrigidity of the machine would have the effect that a force of different amplitude and phase, but of the same frequency, would be applied through the machine holding-down bolts to the inertia block. The possibility of change in amplitude merely adds to the uncertainty already associated with the exciting-force amplitude. Any change in the phase of the force is of no importance in the design of the mounting. The fact that the frequency is unaltered is fortunate because the excitation frequency, in conjunction with the mounting natural frequency and the support natural frequency, is a major design factor.

The procedure given in Chapter 8 for use of the two-mass model does not make direct use of the force amplitude P_S. The mounting is designed to have a nominated (low) value of T_F, so that the transmitted force P_T will be a small fraction, for example, one tenth, of P_S ($T_F = 0.1$), whatever the value of P_S may be. In doing this the uncertainty is, of course, simply transferred to P_T the amplitude of the transmitted force.

This is not a matter of direct concern in practice because, in the present state of the art (see Chapter 5, Performance Testing, Source Mountings), we cannot measure P_T and therefore cannot specify a limiting value as an acceptance criterion. The important quantity is the vibration of the supporting structure that is caused by the transmitted force. All criteria of structural vibration are in terms of the displacement, velocity, acceleration, and frequency (Chapter 4). The calculation of the vibration that a sinusoidal force input at one part of a structure will cause at another part is a matter for the structural engineer versed in structural dynamics and is beyond the scope of this book.

In short, by representing the machine as a lumped mass we are, in effect, adding to the uncertainty of the value of the exciting force P_S. Then, in using the two-mass model to design a mounting to have a specified T_F, the uncertainty attaches to the value of the transmitted force P_T, and we are faced with the question whether the vibration caused by this force will be within acceptable limits.

Excitation at Support of Sensitive Equipment

The two-mass model in Fig. 6.7b does not represent the flexibility of the floor or structure supporting a sensitive equipment installation. The mounting is designed to have a nominated (low) value of T_R which is defined, for sinusoidal excitation, as the ratio of Z_R, the amplitude of the relative displacement between the critical element and the base of the equipment, to Z_S, the displacement amplitude of the support. The excitation frequency of the support, together with the natural frequency of the critical element and that of the mounting, is a major factor in determining the T_R.

If the vibration of the support is from a steady-state source external to the installation as in Fig. 6.7b, the frequency of the vibration of the support is the same as that of the source regardless of the characteristics of the support. However, if the vibration of the support is a transient caused by an impulsive source elsewhere on the site the vibration frequency of the support does depend on the dynamic characteristics of the support; hence, there is good reason for preferring the three-mass model in Fig. 6.9b which would allow the support characteristics to be taken into account.

In adopting the two-mass model the author recognizes this limitation. However, the designer can make some compensation for this inadequacy of the model when the excitation is a transient by checking the design for additional values of the excitation frequency in a range containing an estimated value of what that frequency may be when the site is loaded by the installation.

A further deficiency of the two-mass model is that it involves some uncertainty about the amplitude of the displacement of the support. A value of the amplitude based on measurements at the support before it is loaded by the installation may be unrealistic. If subsequently the added mass of the installation happens to "tune" the support to resonance with the excitation frequency, the support displacement amplitude may be appreciably greater than the value assumed in the design of the mounting. Therefore, the designer should consider the possibility of resonance of the support and, if appropriate, assume a higher value of the amplitude and design for a lower value of T_R so that the response of the critical element will be within acceptable limits.

Responses in Directions Other Than Vertical

The two-mass model represents only vertical motion of the system. This, too, is a significant limitation of the effectiveness of the model. Although the two-mass model is applicable also to unidirectional horizontally vibrating systems and to the analogous two-inertia rotational system, the need to apply the model in these ways arises only rarely in the design of seismic mountings.

Insofar as the subsystem representing the mounting is concerned, the assumption of purely vertical motion is not as serious a limitation as might at first appear. It does not mean that a set of identical isolators must be positioned symmetrically about the vertical axis through the c.g. of the seismic mass. The requirement is that the resultant of the vertical restoring forces acting at all the isolators should have a line of action that passes through the c.g. This is to ensure that the vertical free vibration of the seismic mass is uncoupled with motion in any other mode of free vibration. In Chapter 7 we see that this requirement can be satisfied when the isolator layout is unsymmetrical, if certain conditions affecting the stiffness and layout of the isolators are satisfied.

However, even if the mounting considered separately has an uncoupled vertical mode of free vibration, when it is part of the two-mass system it could be caused to vibrate in other modes because of excitation in directions other than the vertical through the c.g. For this reason the designer must try to arrange for the natural frequencies of the mounting in its various modes to be well separated from the frequencies associated with the excitation.

There is no difficulty in calculating the six natural frequencies of a seismic mass on isolators using the method given in Chapter 7, which treats the mounting alone as a one-mass system with six degrees of freedom. The natural frequency calculations become more complicated when the mounting system is coupled to a second mass–spring system, and the results show that the natural frequencies of coupled systems are not the same as those of the systems considered individually.

For a sensitive-equipment installation, usually the equivalent mass of a critical element is a small fraction of the seismic mass of the mounting and hence the effect of coupling can be ignored. However, for a source mounting the equivalent mass of the support may be comparable with the seismic mass and hence the effect of coupling appreciable. An illustration of the effect of coupling is given in Chapter 8 in relation to the vertical natural frequencies of the two-mass system.

Nonlinearity of Stiffness and Damping

The two-mass model with linear stiffness and viscous damping ignores the nonlinearity of actual isolators. In the practical design of mountings, nonlinearity of stiffness can be ignored because the vibratory displacement of the seismic mass in service is normally small in comparison with the displacement over which the nonlinearity of the stiffness is significant. Therefore, the stiffness of the mounting can be "linearized" at the value appropriate to the working conditions, in the manner discussed in relation to rubber isolators in Chapter 5, and illustrated in Chapter 7 (Example 7.2).

The damping is of secondary importance. A numerical value is seldom specified; the designer simply specifies a type of isolator that will provide damping that is minimal under service conditions, yet adequate to limit the response of a source mounting under resonance conditions, which commonly occur during starting up and slowing down, or to suppress the free vibrations of a sensitive-equipment mounting.

In the calculation of natural frequencies, the damping is ignored because the amount of damping normally involved is too small to have a significant influence on the natural frequencies.

For these reasons the characteristics of the damping are unimportant, assuming that any nonlinearity is not of an extreme kind, for example, dry frictional damping requiring a high threshold force to be overcome before relative motion can occur.

Lumped or Distributed Properties

The two-mass model ignores the fact that the mass, stiffness, and damping properties of actual systems are distributed, not lumped. We have already discussed the use of a lumped mass to represent a machine. For the mounting, the use of a lumped seismic mass is usually defensible. Good practice demands that the inertia block be made stiff enough to ensure that the flexural vibration is negligible. However, flexural vibration is important when the seismic mass is long in comparison with its width and depth, as discussed in Chapter 5 under Design Guide, Stiffness of Inertia Block.

It is not so easy to justify the use of lumped mass, stiffness and damping to represent the floor in Fig. 6.6 or the critical element in Fig. 6.7. The advantage in using lumped properties is that the theoretical analysis is easier than with distributed properties, and the penalty is that the lumped system represents the behavior of the floor or the critical element in only one mode.

The use of a lumped equivalent system for the suspended floor means that the designer is obliged to find design values for the natural frequency of the floor in its fundamental mode and the stiffness of the floor where the mounting is installed. Today these quantities can be calculated from drawings, and for existing floors can be found experimentally. Likewise, the use of a lumped system for the critical element of a sensitive equipment means that the designer must determine the stiffness and natural frequency of the critical element in order to assign lumped properties to the system representing that element in the model.

The properties of the isolators are represented by lumped values of stiffness and damping purely in the interests of simplicity. The behavior of isolators taking into account their distributed mass, and consequently the effects of surging or "wave effects," is reviewed by Snowdon (1979, pp. 33–37). In practice these effects are likely to be important only with high frequency excitation (hundreds of hertz).

Summing Up

The two-mass model as presented in this book is the simplest that can represent the influence of the support in transmitting vibration from a machine installation (Fig. 6.6), or the influence of the critical parts of a sensitive equipment subjected to site vibration (Fig. 6.7.).

The limitations of the two-mass model have been pointed out, the two more serious being that the model represents only unidirectional modes of vibration and does not represent the influence of the support of a sensitive equipment installation. The effects of some of these limitations can be minimized by applying the design data for the two-mass model in ways indicated in Chapters 8 and 9 to cover uncertainties in the values of particular design parameters. Other important considerations, such as the possibility of resonance in modes other than the unidirectional, must be investigated independently of the use of the two-mass model.

Having in mind that the level of general practice in the design of seismic mountings seems to be that of the damped one-mass model, the author considers that a better purpose is served at present in explaining and stimulating the practical application of the two-mass model than in developing design data for more refined models.

PRACTICAL APPLICATION OF TWO-MASS MODEL

We now indicate the specific ways in which the two-mass model is applied in the design of mountings for the isolation of periodic vibration. We also make some general comments on the application of the two-mass model with other basic types of excitation.

Periodic Excitation

Theoretical analysis of the innocent-looking two-mass system to determine its transmissibility is surprisingly more complicated than that of the one-mass system. The latter involves three variables: the excitation frequency, and the natural frequency and damping of the mounting. The two-mass system involves these same three plus the natural frequency and damping of the secondary system (Fig. 6.6 or 6.7) and the ratio of the two masses.

Because the algebraic expressions for the transmissibility ratio turn out to be too lengthy and cumbersome for convenient direct use in the design office, the author and colleagues have evaluated them numerically for combinations of the variables in ranges met in normal practice, and collated the results into graphic data suitable for use in the design of mountings. The reader will see in later chapters that the design data presented in this way give a clear picture of the influence of the flexibility of the support of a source mounting, and of the influence of a critical element in the design of a mounting for a sensitive equipment.

Results derived for the two-mass model give relationships among design variables associated with the mounting, the equipment, and the excitation. Some combinations of these variables promise good isolation; others threaten a magnification of the vibration. In principle these results, if applied early enough in the planning and design of a building, could influence the choice of type of machinery and its operating speeds, the choice of sensitive equipment, and the design of the floors, as well as the design of the mountings. However, in practice, usually all the variables except those of the seismic mounting are determined by factors other than vibration. In other words the data, despite their broader potential applications, will probably be used only to find what should be the natural frequency and damping of the mounting given the values of the other design variables. The task for the designer of the mounting is to select isolators and arrange them so that the mounting will have the required natural frequencies.

Random Excitation

It may be recalled from the discussion in relation to Fig. 2.7 that random vibration contains energy at all frequencies in a certain frequency band, and hence can cause resonance of parts of structures and equipment having natural frequencies in that frequency band. Thus, a machine that is a source of random vibration, for example, a jet engine, could cause resonance of its mounting and/or support. Random vibration at the site of a sensitive equipment could cause resonance of the mounting or of a critical element of the equipment.

The design objective is to reduce to an acceptable level the rms value of the force transmitted from source into support, or to reduce the rms level of the response of the critical elements of a sensitive equipment to random vibration of its support.

The considerations involved in the choice of the design model are basically the same as those discussed in relation to periodic vibration, but the mathematical analysis of the response to random excitation is more difficult (e.g., see Curtis and Boykin, 1961) and is beyond the scope of this book.

Shock Excitation

Perhaps the most familiar shock source in industry is the forging hammer. A hammer blow applied to the worked metal imparts an impulse to the anvil, which is attached to a massive reinforced concrete block resting on isolators. The objective in the design of the seismic mass and choice of the isolators is to minimize the peak force transmitted to the support, by allowing an appreciable downward movement of the seismic mass without significantly reducing the effectiveness of the hammer blow and without objectionable subsequent motion of the hammer as a whole.

The two-mass model can be used as a basis of analysis. The impulse, which is a force pulse of very short duration (e.g., some milliseconds) is idealized as a half-sine pulse, and expressions are derived for the force transmitted and the displacement of the seismic mass (e.g., Crede, 1951, pp. 122–135). The design of mountings for forging hammers is a matter for the specialist. If, as is usual, resilient material is provided between the anvil and the inertia block to reduce the shock input to the block, the analysis calls for the use of a three-mass model. The inertia block must be designed to withstand the stressing produced by the shock waves initiated by the hammer blow. For the theory and practice of shock isolation in general, reference may be made to Crede (1951, Chaps. 3 and 6), Barkan (1962, Chap. 5), and to Newton (1976, Chap. 31).

Transient Excitation at Support of Sensitive Equipment

Shock excitation that originates as an impulse or force pulse is modified by the mass, elasticity, and damping of the material through which the energy is transmitted (Chapter 3). Consequently, the operation of a shock source such as a forging hammer or a metal stamping or shearing press causes the ground and structures in the vicinity to respond with transient vibrations. In the application of the two-mass model in Fig. 6.7, this excitation may be idealized as a damped harmonic displacement of the support, the characteristics of which are influenced by the properties of the support. The application of the two-mass model to the design of mountings for sensitive equipment subjected to a damped harmonic displacement of the support is discussed in Chapter 9.

EXPERIMENTAL VERIFICATION OF MODEL VALIDITY

Recalling the caution given when discussing the role of a design model that one should not put too much trust in a model until its adequacy and validity have been confirmed by experiment, the reader may well ask if we have such confirmation for the two-mass models in Fig. 6.6b and Fig. 6.7b. We have not. The data derived theoretically for these models have only recently been completed and published, and now await application.

Experimental confirmation can be thought of in two ways: confirmation that the data are valid for hardware that, for the purposes of the experiment, has been made to represent the model; and confirmation that the data are valid for actual installations.

A systematic and comprehensive experimental confirmation of the first kind would require the facilities of a laboratory or test house capable of setting up and testing a reasonably sized mock-up of a particular type of installation. This would make use of lumped masses in the form of concrete blocks in a suitable range of sizes, and isolators having linear stiffness and viscous damping in suitable ranges. Facilities would be required for applying suitable test excitations of variable frequency and amplitude, and instrumentation and techniques for measuring transmissibility ratio.

All this adds up to a lengthy and costly test program, which has not been undertaken. It could well be argued that an experiment of this kind is pointless because knowledge of the behavior of unidirectional, linear, two-mass systems is so well established that if a test installation were made having exactly the idealized properties defined in the model, its behavior must agree, within the limits of experimental error, with that derived theoretically for the model. In other words, any disagreement greater than that associated with the experimental uncertainties would be attributable to failure to build into the experimental hardware the idealized properties of the model.

The challenge in practice is to exercise judgment and experience in applying the model to installations that are not tailored to fit the model. Hence, it seems preferable to undertake experimental studies of the second kind mentioned earlier: testing to determine the validity of the two-mass model in relation to actual installations.

Unfortunately, the testing of seismic mountings is rather specialised and costly, and hence is rarely done to satisfy any curiosity or anxiety on the part of the designer about the validity of the design model. Usually, a vibration investigation of a completed installation is done only if the performance is unsatisfactory. The theoretical basis and practical difficulties involved in the performance testing of seismic mountings are discussed in Chapter 5.

The author and colleagues have made vibration measurements on many installations on seismic mountings. It was the investigation of unsatisfactory behavior of some of these that drew attention to the importance of the support as a factor influencing the design of a source mounting, and the significance of relative motion of parts of a sensitive equipment as a factor in the design of its mounting.

Those interested and concerned in the design of seismic mountings are invited and encouraged to take any opportunity to test the performance of mountings, and thereby to build up a background of experience of the extent to which the two-mass model is reliable for particular kinds of vibration-source and sensitive-equipment installation. Incidentally, such testing would no doubt disclose the need to refine the model and derive new design data for some types of installation.

REFERENCES

Barkan, D. D. (1962). *Dynamics of Bases and Foundations*, McGraw-Hill, New York.

Bishop, R. E. D. (1979). *Vibration*, 2nd ed., Cambridge University Press, Cambridge, England.

Crede, Charles E. (1951). *Vibration and Shock Isolation*, Wiley, New York.

Curtis, A. J., and T. R. Boykin, Jr. (1961). Response of Two-degree-of-freedom Systems to White Noise Base Excitation, *J. Acoust. Soc. Am.*, **33**(5), 655–663.

Den Hartog, J. P. (1956). *Mechanical Vibrations*, 4th ed., McGraw-Hill, New York.

Hager, R. W. (1980). Dynamic Analysis and Design—Challenge for the Future, *Shock Vibration Digest*, **12**(4), 3–12.

Macinante, J. A., B. Dorien-Brown, J. L. Goldberg, N. H. Clark, R. A. Glazier, and K. M. O'Toole (1967). A Vibration Study of the C.S.I.R.O. 210-ft Radio Telescope, *Inst. Mech. Eng. London*, Mech. Eng. Sci. Monograph No. 6.

Macinante, J. A. (1969). Design Model Based on Observed Modes of Vibration of Australian C.S.I.R.O. 210-ft Radio Telescope, *Shock Vibration Bull.*, U.S. Naval Research Lab, **40**(4), 155–162.

Newton, R. E. (1976). Theory of Shock Isolation, in *Shock and Vibration Handbook*, 2nd ed., Chap. 31, Cyril M. Harris and Charles E. Crede, Eds., McGraw-Hill, New York.

Ruzicka, Jerome E., and Richard D. Cavanaugh (1958). Elastically Supported Damper System Provides a New Method for Vibration Isolation, *Machine Design*, **30**(21), 114–121.

Ruzicka, Jerome E., and Thomas F. Derby (1971). *Influence of Damping in Vibration Isolation*, Shock and Vibration Monograph SVM-7, Shock and Vibration Information Center, United States Department of Defense, Washington, D.C.

Snowdon, John C. (1979). *Vibration Isolation: Use and Characterization*, U.S. Department of Commerce, National Bureau of Standards, Handbook 128.

7

Free or Natural Vibrations

In the behaviour of this spring-supported mass there is something almost human; it hates to be rushed. If coaxed gently, and not un-naturally hurried, it comes with perfect docility in the direction in which it is being urged; but, if asked to bestir itself at more than its customary gait, it displays a mulish perversity of disposition. Such motion as it can be prevailed upon to make is in a retrograde direction, and the more it is rushed the less it condescends to move. But, if it is stimulated by a force having its own inherent natural frequency, it shows its delight by bouncing up and down with an exuberance of spirit which may be very embarrassing.

CHARLES EDWARD INGLIS (1944, pp. 316, 318)

Any beam, column, floor, or other material system having the properties of mass and elasticity is capable of making a free or natural vibration. If the system is displaced from its rest or unstrained configuration, elastic strain introduces internal forces and moments which oppose the strain. For brevity and by analogy with the simple mass-spring system we shall refer to the effects of this elastic strain as a "restoring force." If, then, the system is suddenly released, the restoring force accelerates the system back toward its rest position, but by the time it has arrived there it has acquired momentum and overshoots the rest position. This brings into action a restoring force in the opposite direction which resists the overshoot, brings the system momentarily to rest, and immediately accelerates it back again through the rest position, and so on. This free vibration goes on until the system is brought to rest by damping which dissipates the energy of the free vibration.

The frequency of this free vibration is the "inherent natural frequency" referred to in the opening quotation. Its precise value varies to some extent with the damping and amplitude, but for the systems and applications considered in this book we can assume that the natural frequency is constant for a given system and mode of vibration.

The natural frequency of a spring-supported mass is the key to its behavior when acted on by an oscillatory force. The simple experiment illustrated in Fig. 6.2 shows the three kinds of response described in the quotation: the in-phase motion of the mass when the forcing frequency is well below the natural frequency, the large response at resonance when the forcing frequency equals the natural frequency, and the small antiphase response when the forcing frequency is well above the natural frequency.

Resonance occurs also in acoustical, electrical, and other physical systems. For example, in the tubes and pipes of a variety of musical instruments a column of air is set into resonance to make the note; a radio receiver is tuned by an adjustment that makes its electrical natural frequency equal to the carrier frequency of the signal transmitted from the radio station; resonances that occur in molecules have important applications in research and technology.

In the design of seismic mountings we are concerned only with mechanical resonances of the unwanted kind that can occur in machine and equipment installations. In this context our interest in resonance is like our interest in income tax or the common cold; we study it so we can learn how to avoid, or at least minimize, its effects.

In the discussion so far we have implied that the mass–spring system has one mode of vibration and hence one natural frequency. A seismic mounting is a mass on springs that is not so constrained. It is capable of making natural vibrations in six modes and therefore of having unwanted resonances in six modes. These are the six modes of free vibration of a rigid or lumped mass illustrated in Fig. 2.10.

During a free vibration of a mass on isolators the inertial effects associated with acceleration and deceleration are distributed throughout the mass, with the result that elastic modes of vibration of the kinds discussed in relation to Fig. 2.11 and Fig. 2.12 are superposed on the rigid-body modes of Fig. 2.10. In the design of seismic mountings the elastic vibrations are ignored unless the inertia block is long in comparison with its depth or width (e.g., see Example 5.9) in which case the flexural and torsional modes should be checked and the inertia block designed to ensure that its natural frequencies in these elastic modes are well separated from the excitation frequencies.

In this chapter we assume that the seismic mass is rigid; therefore, we can represent the inertial effects by a single or total inertial force acting at the c.g. of the mass, and we deal only with the six rigid-body modes.

In the spirit of the opening quotation, the designer of a mounting has two objectives: to frustrate any exuberant response by making sure that the inherent natural frequency in any mode does not coincide with a forcing frequency that could excite resonance in that mode; and to provoke a high degree of "mulish perversity" by making the ratio of forcing to natural frequency in any mode as high as practicable.

In practice, the forcing frequency is fixed nearly always by existing conditions, or is determined by factors outside the control of the designer of the mounting. Therefore, a favorably high frequency ratio can be achieved only by designing the mounting to have natural frequencies suitably low in comparison with the forcing frequency.

To do this the designer must have a clear picture of the kinds of natural vibration that can occur when a spring-supported mass is disturbed then allowed to vibrate freely, and must know how to calculate the natural frequencies in terms of the hardware: specifically the shape and size of the seismic mass, the properties of the isolators, and their positions and orientation.

For the calculation of natural frequencies, those who are well armed mathematically and favor a general approach like to begin by considering a seismic mass of arbitrary size and shape supported on an assortment of springs of different stiffness, arranged without symmetry of layout or uniformity of orientation. If such a system is disturbed then allowed to vibrate freely its motion is indescribable in simple and meaningful prose, for there is no sustained motion in any one mode. All six modes are coupled and the motion can be described only mathematically, by using a set of six simultaneous equations. This highly generalized approach is more noteworthy for its academic quality than its usefulness to the practitioner.

In this book the treatment is no more general than is necessary for the design of mountings of the types commonly used in stationary installations. The mounting is designed to have as many uncoupled modes as practicable. An uncoupled mode is one in which the system can vibrate freely, independently of motion in any other mode. Limitation of the degree of coupling not only simplifies the calculation and adjustment of the natural frequencies, but also enables the designer to visualize more clearly the modes of vibration of the mounting and hence to predict with more confidence its response to excitation.

First, we stipulate that the mounting should have an uncoupled vertical mode. This presents no difficulty. In fact, as we show in the next section, it results automatically from what most designers would do intuitively: they would use isolators all of the same stiffness and load rating, and arrange them to share the load equally. We see later that provision of an uncoupled vertical mode yields also an uncoupled rotational mode about the vertical axis. We then consider the coupled and uncoupled horizontal and rotational modes, and derive the conditions for minimal coupling in a base-type mounting. In all the discussion relating to the natural frequencies of a seismic mounting, we assume that the isolator positions are defined by their coordinates referred to a set of rectangular axes having their origin at the c.g. of the seismic mass in its rest position, and that the principal inertial axes of the seismic mass do not deviate significantly from the coordinate axes.

VERTICAL MODE

In the course of a vertical free vibration, the inertial force acts vertically through the c.g. and is opposed by restoring forces at all the isolators. In Fig. 7.1 the resultant (R) of all these individual restoring forces is assumed to be in the xz plane and to the right of the inertial force (F). When the c.g. is below its rest position O, as in Fig. 7.1, the two forces form a couple that tilts the mass anticlockwise, and when the c.g. is above O the inertial force acts upward, the restoring force downward, and the mass tilts clockwise. It is not obvious that the restoring force can act downward if the isolators remain in compression throughout the cycle: the physical explanation of this is given later, in the context relating to Fig. 7.5. Thus, the vertical-z mode is coupled with rotation about Oy. If the restoring force R is not in the xz plane, vertical translation is coupled also with rotation about Ox.

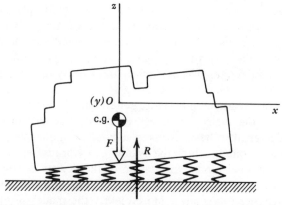

Figure 7.1.   Vertical motion coupled with rotation about a horizontal axis.

If the mass is to have an uncoupled vertical mode the resultant restoring force must have a line of action, sometimes called the vertical elastic axis of the isolator system, that passes through the c.g. of the mass. For, if it does, the inertial force acting upward and downward through the c.g. is always opposed by a restoring force having this same line of action; therefore, there is no couple tending to tilt the mass. An equivalent statement of the requirement is that throughout the cycle of vertical vibration all the isolators deflect in phase and with the same amplitude, for they can do so only if there is no rotation about a horizontal axis.

Uncoupled Vertical Mode

The requirement for an uncoupled vertical mode is illustrated in Fig. 7.2a, which shows the seismic mass at some instant when the c.g. is below its rest position O. The inertial force F is directed downward and the resultant restoring force R is required to act upward in the same line of action as F.

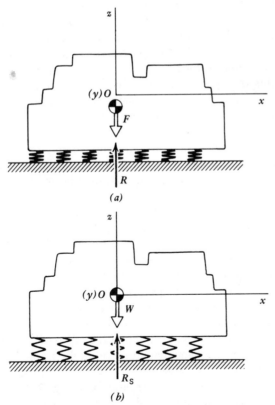

Figure 7.2. (a) Dynamic and (b) static forces in a mounting with uncoupled vertical mode.

In a practical design situation this means that the system of isolators must be designed to have R in the right place, because little can be done to modify the line of action of F by altering the position of the c.g. The latter is determined by the c.g. of the mounted equipment, which is inherent in its design, and that of the inertia block which is fixed by its shape, size, and composition, which in turn are determined by considerations other than that of providing an uncoupled vertical mode.

The line of action of R is determined by the magnitudes and lines of action of the individual restoring forces of which R is the resultant. The magnitude of the restoring force at any one isolator is determined by its stiffness and deflection. In practice normally all isolators in a given mounting are of the same stiffness. If the c.g. of the seismic mass is much closer to one end of the seismic mass than the other, stiffer isolators could be used under the "heavy end" but usually it is convenient to use isolators all of the same stiffness and to provide additional isolators under the heavy end. Therefore, to avoid unnecessary complication, we assume that all isolators used in a given mounting are of nominally the same stiffness and load-carrying capacity; for brevity we shall refer to these as *identical isolators*.

The restoring force at each isolator is proportional to its deflection, and since in a purely vertical vibration the deflection at any instant must be the same at all the isolators, and all the isolators are assumed to have the same stiffness, the individual restoring forces must all be of the same magnitude at any instant during the cycle. Therefore, in designing a mounting on identical isolators to have an uncoupled vertical mode, the isolator layout must be such that the set of equal restoring forces, one at each isolator, has a resultant that passes through the c.g.

It is perhaps surprising, and certainly convenient, that this requirement concerning the dynamic forces associated with a vertical free vibration can be satisfied simply by arranging for the static load to be shared equally by all the isolators.

A little reflection will show why this is so. Referring to Fig. 7.2b, which shows the system at rest, if the weight W of the seismic mass is balanced by a set of equal static forces at the isolators, then the resultant R_s of these static forces must pass through the c.g. because if it did not the mass would tilt and the isolators would not have equal deflection and hence would not share the load equally. Now if the isolator positions are such that a certain force [the static force W (Fig. 7.2b)], acting vertically through the c.g., produces equal static deflection of all the isolators, then another force [the dynamic force F (Fig. 7.2a)], also acting vertically through the c.g., will produce deflection that at any instant is of the same magnitude at all the isolators, and this is the requirement for an uncoupled vertical mode. In an uncoupled vertical vibration the static and dynamic forces shown separately in Fig. 7.2 are superposed.

In the foregoing discussion no reference has been made to the load/deflection characteristics of the isolators. The statement that the dynamic requirement for an uncoupled vertical mode is satisfied by arranging for equal sharing of the static load is applicable to nonlinear as well as linear isolators, provided that the isolators have identical load/deflection characteristics.

How does the designer arrange the positions of the isolators so that they share the static load equally? Let us assume that we have a preliminary design for a base mounting involving a layout of a certain number of identical isolators numbered $1, 2, \ldots, n$ in Fig. 7.3. Although the isolators are on a plane some distance below the xy plane through the c.g., for convenience we define their positions by their coordinates referred or projected to the xy plane, as shown.

The object is to find the line of action of the resultant (R_s) of a system of equal vertical forces, one at each isolator so that if, as in Fig. 7.3, the line of action does not pass through the c.g., the designer can alter the isolator positions so that it does. The method of doing this (Macinante, 1960) is as follows.

It can be shown from elementary statics that a set of n equal vertical forces whose lines of action are defined by the coordinates x_1,y_1; x_2,y_2; $\ldots$; x_n,y_n as in Fig. 7.3 has a resultant whose line of action passes through the point x_R,y_R, where

$$x_R = \frac{1}{n} (x_1 + x_2 + \cdots + x_n)$$

$$y_R = \frac{1}{n} (y_1 + y_2 + \cdots + y_n)$$

By analogy one can think of the point x_R,y_R as the c.g. of a set of equal masses located at the isolator positions.

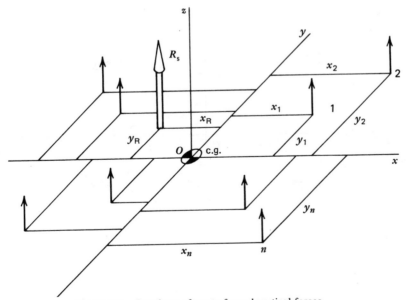

Figure 7.3. Resultant of a set of equal vertical forces.

Since the c.g. of the seismic mass is taken to be the origin of the coordinates, the resultant R_s is on the vertical axis Oz if $x_R = y_R = 0$, that is, if the algebraic sums of the x- and y-coordinates are both zero:

$$x_1 + x_2 + \cdots + x_n = 0$$

$$y_1 + y_2 + \cdots + y_n = 0$$

(7.1)

If a nominated layout of isolators is such that the sums of the coordinates are not zero, then the positions of some or all of the isolators should be adjusted to make the sums zero, as illustrated in the following example:

Example 7.1.   A preliminary design for a seismic mounting proposes the use of twelve isolators all of the same stiffness and load rating, in the layout shown in Fig. 7.4 and detailed in the x and y columns in Table 7.1. (Such a haphazard arrangement of isolators is unlikely to occur in practice; it is proposed only for the purpose of this example.) Check and if necessary modify the layout to provide an uncoupled vertical mode.

By adding the entries in the x and y columns in Table 7.1 it is found that the algebraic sum in each column is not zero. The positions of the isolators could be modified in many different ways to make these sums zero. One of these ways is to move isolators numbered 3, 7, 8, and 10 to the positions

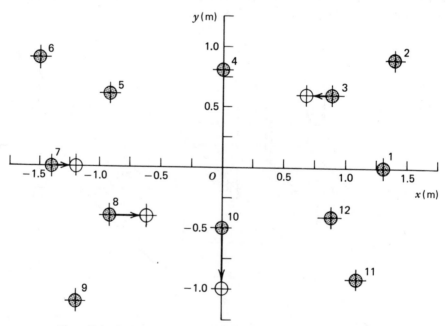

Figure 7.4.   Isolator layout for uncoupled vertical mode (Example 7.1).

Table 7.1. Calculations for Uncoupling of Modes and Natural Frequencies (Examples 7.1, 7.3, 7.5, and 7.7)

Isolator No.	x (m)	y (m)	xy (m^2)	x^2 (m^2)	y^2 (m^2)
1	1.3	0	0	1.69	0
2	1.4	0.9	1.26	1.96	0.81
3	0.9	0.6	0.54	—	0.36
	(0.7)		(0.42)	0.49	
4	0	0.8	0	0	0.64
5	−0.9	0.6	−0.54	0.81	0.36
6	−1.5	0.9	−1.35	2.25	0.81
7	−1.4	0	0	—	0
	(−1.2)		(0)	1.44	
8	−0.9	−0.4	0.36	—	0.16
	(−0.6)		(0.24)	0.36	
9	−1.2	−1.1	1.32	1.44	1.21
10	0	−0.5	0	0	—
		(−1.0)	(0)		1.0
11	1.1	−0.9	−0.99	1.21	0.81
12	0.9	−0.4	−0.36	0.81	0.16
Sum	−0.3	+0.5	+0.24	12.46	6.32
	(0)	(0)	(0)		

shown in Fig. 7.4 and in parentheses in Table 7.1. The modified layout satisfies the requirements for an uncoupled vertical mode.

Table 7.1 includes columns for xy, x^2, and y^2. These are required in the solution of examples to follow.

Natural Frequency in Uncoupled Vertical Mode

Consider a seismic mounting on any number of identical isolators, linear or nonlinear, with their loading axes vertical, in any layout, symmetrical or unsymmetrical, that satisfies the condition [Eq. (7.1)] for equal sharing of the static load.

During a vertical vibration the deflection of any isolator at any instant is made up of a static and a dynamic component. With the seismic mass at rest any one isolator, shown as a helical spring in Fig. 7.5a, carries its share (w) of the total "dead weight" W of the seismic mass, under which its deflection from the unloaded condition is the static deflection $AB = z_s$.

The vertical vibratory deflection takes place above and below the rest or datum position B. Figure 7.5b shows a cycle of the deflection/time curve, which is the same for all the isolators, and also represents the vertical displacement of the c.g. of the seismic mass.

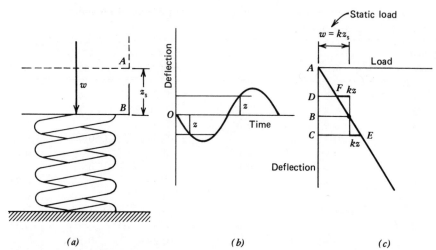

Figure 7.5. Static and dynamic deflections of isolator during uncoupled vertical vibration.

The discussion at present refers to helical spring isolators, which have a linear load/deflection relationship. Later we show how the results are applied to nonlinear isolators. The load/deflection relationship for one spring isolator is shown in Fig. 7.5c, and is the same for all the isolators.

The static load on each isolator is equal to the total load divided by the number of isolators. The number is decided by taking into consideration the stiffness and load rating of available isolators, and any limitations on the places on the floor or seismic mass where isolators can be accommodated.

The static load on each isolator is given by

$$w = \frac{W}{n} = kz_s$$

where w = static load on each isolator (N)
$\quad W = Mg$ = weight of seismic mass (N)
$\quad M$ = seismic mass (kg)
$\quad g$ = acceleration of gravity (m/s^2)
$\quad n$ = number of isolators
$\quad k$ = stiffness of isolator (N/m)
$\quad z_s$ = static deflection of isolator (m)

During a vertical free vibration, the compressive force in each isolator is alternately greater and less than the initial constant compressive force w. Referring to Fig. 7.5b and 7.5c, when the mass is instantaneously some distance z below the datum, the compressive force CE in the isolator is greater than the gravitational force w by the amount kz, which therefore represents the upward restoring force. When the mass is some distance z above the datum, the compressive force DF in the isolator is less than w by the amount kz, which means that the downward restoring force results from the fact that w is greater than the compressive force in the isolator. It can be seen in the diagram that the net downward restoring force is kz.

The foregoing discussion is applicable also to a mounting on a set of identical nonlinear isolators. This can be seen by following the same line of discussion but substituting for the linear static deflection (z_s, Fig. 7.5) the nonlinear static deflection (d, Fig. 5.8), and substituting for the restoring force (kz) of a linear isolator the value ($k_d z$), where k_d is the dynamic stiffness of the nonlinear isolator as discussed in Chapter 5 under the heading Rubber Isolators.

The natural frequency in an uncoupled vertical mode can be calculated from

$$f_z = \frac{1}{2\pi} \left(\frac{K_z}{M} \right)^{1/2} \tag{7.2}$$

where f_z = natural frequency in vertical (z) direction (Hz)
   K_z = nk_z = total dynamic stiffness in vertical (z) direction (N/m)
   k_z = dynamic stiffness of one isolator in vertical direction under the conditions of vibration experienced by the isolator in the particular application (N/m)
   n = number of isolators
   M = seismic mass (kg)

The derivation of Eq. (7.2) can be found in vibration textbooks (e.g., Den Hartog, 1956, pp. 31–34). It involves the assumptions that the damping is too small to have a significant influence on the natural frequency, and that the force/displacement relationship is linear in the range of the vibratory displacement concerned.

The expression for the natural frequency is commonly written in a simpler form by making use of the fact that the static deflection (z_s) of a linear spring system of stiffness K_z is proportional to the load. Thus,

$$z_s = \frac{Mg}{K_z} \quad \text{or} \quad \frac{K_z}{M} = \frac{g}{z_s} \tag{7.3}$$

For a helical spring isolator it can be assumed that the dynamic and static stiffness are equal; therefore, Eq. (7.3) can be substituted into Eq. (7.2) with the result:

$$f_z = \frac{1}{2\pi} \left(\frac{g}{z_s} \right)^{1/2}$$

By substituting $g = 9.8$ m/s^2 and expressing z_s in millimeters, this becomes

$$f_z \approx 16 \left(\frac{1}{z_s} \right)^{1/2} \tag{7.4}$$

This simple formula for calculating the natural frequency has found widespread use in the practical design of mountings. The suppliers of isolators almost invariably publish graphs, similar to that in Fig. 7.6, from which the natural frequency can be read directly for a given static deflection.

In the beguiling simplicity of this relationship between static deflection and natural frequency lies a trap for the unwary. The relationship is valid only for isolators having a linear load/deflection characteristic not merely in the range of the vibratory displacement but over the full static deflection under the working load.

Example 7.2.   Calculate the vertical natural frequency of a seismic mounting in which a total seismic mass of 36 tonnes rests on a set of 12 identical rubber-in-shear isolators in the layout arrived at in Example 7.1 to provide an uncoupled vertical mode. The static load/deflection curve for one isolator is given in Fig. 7.7. From technical data provided by the suppliers of the isolators, the ratio of dynamic to static stiffness is 1.6.

The vertical mode of the mounting in Example 7.1 was arranged to be uncoupled by modifying the isolator coordinates so that they satisfy Eq. (7.1) and hence they share the static load equally. Therefore, the load per isolator is 36,000/12 = 3000 kg.

Referring to the point P corresponding to this load in Fig. 7.7, the static deflection is read as 5.9 mm. By drawing a tangent to the curve at the point

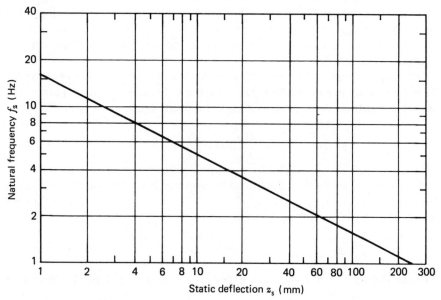

Figure 7.6. Relationship between natural frequency and static deflection, applicable only to isolators having linear load/deflection characteristics.

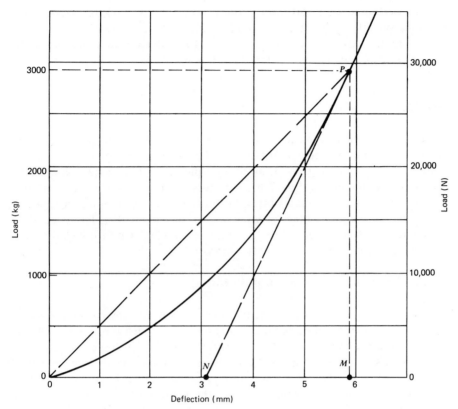

Figure 7.7.   Load/deflection and static stiffness of nonlinear isolator (Example 7.2).

P corresponding to the working load, the slope of the curve is found to be PM/NM = 3000 × 9.8/2.8 = 10.5 × 10³ N/mm. This is the static stiffness (k_s) of one isolator "linearized" in a small range of displacement about the rest position. The dynamic stiffness of each isolator is then k_d = 1.6 k_s = 16.8 × 10³ N/mm = 16.8 × 10⁶ N/m or 16.8 MN/m.

By substituting into Eq. (7.2) the value for the total vertical stiffness K_z = 12 × 16.8 × 10⁶ N/m, and M = 36 × 10³ kg, the natural frequency is found to be 11.9 Hz.

If the difference between the dynamic and the static stiffness were overlooked or ignored, taking the stiffness of one isolator to be 10.5 × 10³ N/mm, the natural frequency would be calculated to be 9.4 Hz.

If the simple formula in Eq. (7.4) were used, incorrectly treating the nonlinear static deflection z_s = 5.9 mm as though it were the static deflection of a linear system, the natural frequency would be calculated to be 6.6 Hz (see also Fig. 7.6).

Serious trouble can result from incorrectly calculating the natural frequency and subsequently finding that the actual value (e.g., 11.9 Hz in the example) is much higher than expected (e.g., 6.6 or 9.4 Hz) and perhaps embarrassingly close to a resonance frequency (see also Example 5.5).

HORIZONTAL MODES

We now discuss the way in which a free vibration involving horizontal translation of the c.g. can be coupled with rotation, and how the mounting can be designed to minimize the coupling. We shall discuss the coupling of x-translation with rotation about the y-axis (Fig. 7.8), with rotation about the vertical axis, and with rotation about the x-axis of translation. These are the motions commonly called, respectively, pitching, yawing, and rolling (Fig. 2.10). The discussion and conclusions with reference to rotations coupled with x-translation are applicable to rotations coupled with y-translation.

Coupled Pitching and Horizontal Motion

Consider a free vibration that occurs after the seismic mass in Fig. 7.8 has been displaced in the horizontal-x direction then suddenly released. During an excursion to the right of the rest position, the inertial force F is directed to the right through the c.g., and is opposed by a horizontal restoring force R, which is the resultant of the horizontal restoring forces in all the isolators. Because the isolators are below the c.g. the restoring force and the inertial force form a couple, which tilts the mass alternately clockwise and anticlockwise as the c.g. of the mass moves to right and left of its rest position. That is, the x-translation is coupled with rotation about Oy. The natural frequencies in the modes resulting from this coupling are discussed in detail later under the heading Coupled Modes of a Base-Type Mounting.

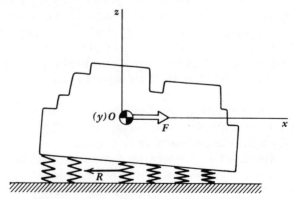

Figure 7.8. Coupled pitching and horizontal motion.

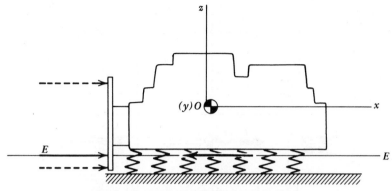

Figure 7.9.   Horizontal elastic axis of isolators.

How can these two modes be uncoupled? To answer this, let us imagine that an end plate is fixed to the mass as shown in Fig. 7.9 so that an external force parallel to Ox can be applied in the xz plane at any desired height in relation to the isolators. A force near the top of the bracket will cause the mass to tilt clockwise as it is pushed to the right; a force near the bottom, anticlockwise. There is a particular height, shown by the line EE, at which the external force will produce x-translation without rotation about Oy.

A horizontal-x free vibration of the mass can take place without rotation about Oy only if the mounting is designed to have the c.g. of the seismic mass on the line EE. Obviously this is impossible with a base-type mounting, and can be achieved only with a c.g.-type mounting (Figs. 5.3 and 5.4).

Coupled Yawing and Horizontal Motion

Assume that we have arranged to eliminate coupling with rotation about Oy by using c.g. mounting. We now wish to eliminate coupling about Oz. If this desired condition is attained, all isolators must experience the same horizontal-x deflection and, assuming that all isolators have the same horizontal stiffness, all must oppose the translation with the same force (r). The resultant (R) of this set of equal forces must pass through the c.g. as shown in Fig. 7.10. This requirement is satisfied if the sum of the anticlockwise moments about Oz, such as ry_1, ry_2, is equal to the sum of the clockwise moments. Since the forces all have the same magnitude (r) the design requirement is that the algebraic sum of the y coordinates be zero.

Similarly, y-translation is uncoupled with rotation about the vertical axis if the algebraic sum of the x coordinates is zero. These conditions for eliminating coupled rotation about the vertical axis are the same as those for an uncoupled vertical mode [Eq. (7.1)]. In practice the isolators in a c.g. mounting normally have a symmetrical layout, which obviously satisfies these requirements for uncoupled rotation about the vertical axis.

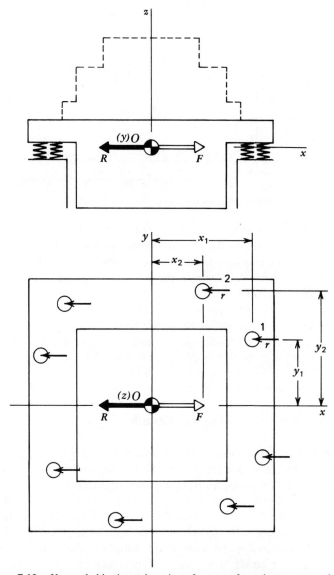

Figure 7.10.   Uncoupled horizontal motion of center-of-gravity type mounting.

We have discussed this with reference to a c.g. mounting. In a base mounting (Fig. 7.8), R and F are not in the same horizontal plane. Nevertheless, if the conditions in Eq. (7.1) are satisfied, both R and F are in the vertical plane xz containing the c.g. and therefore do not form a couple tending to cause rotation about the vertical axis during a free vibration in which the x-translation is coupled with rotation about Oy.

Coupled Rolling and Horizontal Motion

Consider a base mounting on the arrangement of identical isolators shown in Fig. 7.11a, in which, for the purpose of this discussion, most of the isolators in the first and third quadrants are close to the xz plane. For convenience, the isolator positions are referred to the xy plane (Fig. 7.11b).

It can be visualized that in a horizontal-x free vibration, as the mass moves to the right of its rest position as in Fig. 7.8, its tilting clockwise about Oy will be accompanied by some tilting about Ox. Figure 7.11b shows the vertical restoring forces at four representative isolators, one in each quadrant, that would act if the mass were constrained to tilt through a small angle β about Oy without rotation about Ox. The deflection of isolator No. 4 increases by an amount βx_4 and hence its upward restoring force increases by $k_z \beta x_4$. At the

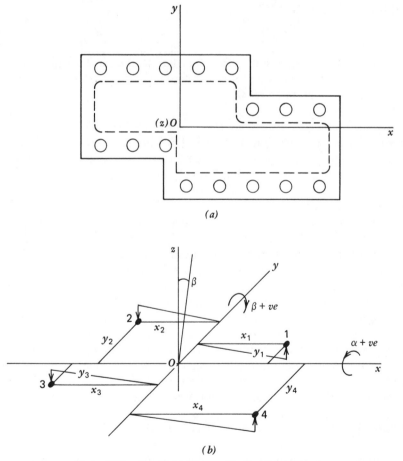

(a)

(b)

Figure 7.11. Coupled rolling and horizontal motion.

same time the downward restoring force at isolator No. 2 increases by $k_z \beta x_2$. The moments of these forces tending to rotate the mass in the α-negative sense about the Ox axis are, respectively, $k_z \beta x_4 y_4$ and $k_z \beta x_2 y_2$. Similarly, isolators Nos. 1 and 3 tend to rotate the mass in the α-positive direction about Ox when it tilts in the β-positive direction about Oy.

By applying to the coordinates of the isolators the conventional algebraic signs, it will be seen that for isolators in the first and third quadrants the xy product is positive and tends to cause α-positive rotation about Ox, whereas those in the second and fourth quadrants have negative xy products and tend to cause α-negative rotation. If the sum about Ox of moments of all the changes in restoring force resulting from the tilting about Oy is zero then there is no tendency of the mass to rotate about Ox. The algebraic sum of these moments is zero if that of the products of the isolator coordinates is zero. In short, the condition for coupled x-translation and pitching about Oy without "cross-coupled" rolling about Ox is

$$x_1 y_1 + x_2 y_2 + \ldots + x_n y_n = 0 \qquad (7.5)$$

If we had begun by assuming a y-translation, the same result would have been derived as the condition for y-translation coupled with rotation about Ox without cross-coupling about Oy.

Example 7.3.   The seismic mounting in Example 7.1, being a base mounting, unavoidably has x-translation coupled with rotation about Oy, and y-translation coupled with rotation about Ox. It is required that in these modes of vibration there should be no "cross-coupling" involving rotation about the axis of translation. Check and if necessary adjust the isolator positions originally proposed, to eliminate cross-coupling.

The unwanted cross-coupling is eliminated if the algebraic sum of the products of the isolator coordinates is zero [Eq. (7.5)]. The solution given to Example 7.1 was chosen in anticipation of the present example. It can be seen in Table 7.1 that the positions of the isolators have been modified so that not only is the algebraic sum of the x- and of the y-coordinates zero, but also that of the xy products is zero, thereby eliminating cross-coupling.

In adjusting the isolator positions, note that isolators in the first and third quadrants have xy products of positive sign, and those in the second and fourth quadrants negative. Isolators on an axis can be moved along that axis to balance Eq. (7.1) without affecting the balance of Eq. (7.5). Thus, isolators Nos. 3 and 8 were adjusted to make the xy sum zero, then isolator No. 7 to make the x-sum zero, and isolator No. 10 to make the y-sum zero.

Natural Frequency in Uncoupled Horizontal Mode

The natural frequency in an uncoupled horizontal mode is calculated by using an equation of the same form as Eq. (7.2), and substituting for the stiffness the sum of the horizontal stiffnesses of all the isolators, as shown in the following example.

Example 7.4. Calculate the natural frequency in the horizontal uncoupled mode of the c.g.-type mounting shown in Fig. 7.12. A total seismic mass of 1.2 tonnes rests on four helical springs which have a linear static deflection of 25 mm. The ratio of horizontal to vertical stiffness of the springs is 0.5.

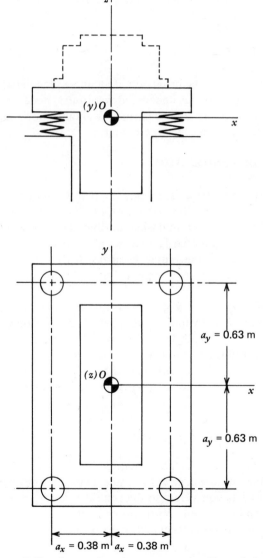

Figure 7.12. Center-of-gravity type mounting (Example 7.4).

Each spring carries a static load of 1200/4 = 300 kg. The vertical stiffness of one spring is $300 \times 9.8/25 = 118$ N/mm, and its horizontal stiffness is $0.5 \times 118 = 59$ N/mm. The total horizontal stiffness of the four springs is $K_x = K_y = 4 \times 59 = 236$ N/mm $= 2.36 \times 10^5$ N/m.

The horizontal natural frequency is found by substituting this value of the total horizontal stiffness, and the value $M = 1200$ kg, into an equation of the same form as Eq. (7.2):

$$f_x = f_y = \frac{1}{2\pi}\left(\frac{2.36 \times 10^5}{1200}\right)^{1/2} = 2.2 \text{ Hz}$$

ROTATIONAL MODES

We begin by showing that the provision of an uncoupled vertical translational mode carries the bonus of an uncoupled rotational mode about the vertical axis. We then consider the rotational modes about a horizontal axis, which can be uncoupled only in a c.g.-type mounting.

Oscillation about Vertical Axis

Consider that a seismic mass, represented in plan view by the solid outline in Fig. 7.13, has rotated about the vertical axis through a small angle γ from its rest position, which is shown in dashed outline. The mass rests on a number of identical isolators, numbered $1, 2, \ldots, n$, at positions having coordinates $x_1,y_1, x_2,y_2, \ldots, x_n,y_n$, and radial distances $d_1,d_2, \ldots, d_n$, respectively, in a layout that satisfies the requirement [Eq. (7.1)] for an uncoupled vertical mode.

The angular displacement γ gives rise to horizontal restoring forces H_1, $H_2, \ldots, H_n$, represented by the arrows at the isolator positions. The horizontal force at each isolator is perpendicular to the radial line to the isolator, and its magnitude is the product of its horizontal deflection $\gamma d_1, \gamma d_2, \ldots, \gamma d_n$ and its horizontal stiffness. We assume the stiffness to have a constant value k_{xy} in any direction in the horizontal plane. Thus, the horizontal forces at the isolators are

$$H_1 = \gamma d_1 k_{xy}, \quad H_2 = \gamma d_2 k_{xy}, \quad \ldots, \quad H_n = \gamma d_n k_{xy} \qquad (7.6)$$

Using a technique of elementary statics, we can consider each of these forces to be equivalent to a central force and a couple. For example, the force H_1 at isolator No. 1 is equivalent to the central force H_1 shown by the arrow in open outline, and the couple $H_1 d_1$ shown by the pair of arrows in full line. Thus, the anticlockwise displacement γ gives rise to a set of central forces

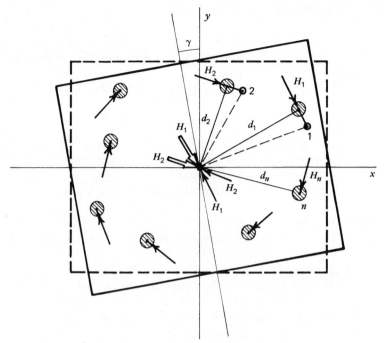

Figure 7.13. Uncoupled rotational mode about vertical axis.

H_1, H_2, . . . and a set of clockwise couples H_1d_1, H_2d_2, ... If the set of central forces is in equilibrium there is no resultant force tending to displace the mass horizontally.

That the set of central forces is in equilibrium if the isolator coordinates satisfy Eq. (7.1) can be demonstrated as follows. The set of forces H_1, H_2, . . . ,H_n concurrent at O comprises one force per isolator, the magnitude of each force being proportional to the radial distance to the isolator [Eq. (7.6)], and its direction perpendicular to the radial line. A vector representation of this set of forces would be geometrically similar to the set of the radial lines d_1, d_2, . . . , d_n rotated anticlockwise through a right angle. The resultant of this set of forces is zero if the algebraic sum of the x-components and that of the y-components of all the forces are both zero. These sums *are* zero if the isolator layout satisfies the requirement [Eq. (7.1)] for an uncoupled vertical mode, because in the vector diagram the lines representing the horizontal and vertical components of the forces are proportional respectively to the x- and y-coordinates of the isolator positions. Consequently, if a mounting is designed to have an uncoupled vertical mode, then also its rotational mode about the vertical axis is uncoupled with horizontal translation.

Natural Frequency about Vertical Axis

The natural frequency in the uncoupled rotational mode about the vertical axis is given by the following equation, which is analogous to that for the uncoupled translational mode [Eq. (7.2)]:

$$f_\gamma = \frac{1}{2\pi} \left(\frac{K_\gamma}{I_z} \right)^{1/2} \tag{7.7}$$

where f_γ = natural frequency in rotational mode about vertical axis (Hz)
  K_γ = angular stiffness, or restoring torque per unit angular displacement about vertical axis (N·m/rad)
  $I_z = M\rho_z^2$ = moment of inertia of seismic mass about vertical axis (kg·m^2)
  M = seismic mass (kg)
  ρ_z = radius of gyration of seismic mass about vertical axis (m)

Some hints are given by Crede (1951, pp. 21–23) on the estimation of radius of gyration in relation to the design of seismic mountings. Formulas for the radius of gyration of bodies of particular shapes are given in handbooks.

The total restoring torque opposing a small angular displacement γ about the Oz axis of the mounting shown schematically in Fig. 7.13 is

$$H_1 d_1 + H_2 d_2 + \ldots + H_n d_n$$

Substituting for $H_1, H_2, \ldots, H_n$ from Eq. (7.6) this becomes

$$\gamma k_{xy} (d_1^2 + d_2^2 + \ldots + d_n^2)$$

Therefore, the restoring torque per unit of angular displacement γ is

$$
\begin{aligned}
K_\gamma &= k_{xy}(d_1^2 + d_2^2 + \ldots + d_n^2) \\
&= k_{xy}(x_1^2 + x_2^2 \ldots + x_n^2 \; ; + \; y_1^2 + y_2^2 + \ldots + y_n^2) \\
&= k_{xy} n (X^2 + Y^2)
\end{aligned}
$$

where X^2 and Y^2 are the mean squared values of the x and y coordinates, respectively:

$$X^2 = \frac{1}{n}(x_1^2 + x_2^2 + \cdots + x_n^2)$$

$$Y^2 = \frac{1}{n}(y_1^2 + y_2^2 + \cdots + y_n^2)$$

By substituting the expression for K_γ into Eq. (7.7) the natural frequency f_γ becomes

$$f_\gamma = \frac{1}{2\pi} \left[\frac{k_{xy}n(X^2 + Y^2)}{I_z} \right]^{1/2} \tag{7.8}$$

Example 7.5. Calculate the natural frequency in the uncoupled rotational mode about the vertical axis of the mounting described in Examples 7.1 and 7.2, taking the layout of the isolators to be that modified in Example 7.1 to give an uncoupled vertical mode. The radius of gyration of the seismic mass about the vertical axis is $\rho_z = 1.2$ m. From information supplied by the manufacturer the isolators have a stiffness ratio $k_{xy}/k_z = 0.6$.

Referring to Eq. (7.8), the various quantities required for the calculation of the natural frequency are evaluated as follows.

From Example 7.2 the vertical dynamic stiffness of one isolator is $k_z = 16.8 \times 10^6$ N/m; therefore, the horizontal stiffness is $k_{xy} = 0.6\ k_z = 10.08 \times 10^6$ N/m.

From Table 7.1 the mean squared values of the isolator coordinates are

$$X^2 = \frac{12.46}{12} = 1.04 \text{ m}^2 \qquad Y^2 = \frac{6.32}{12} = 0.53 \text{ m}^2$$

The moment of inertia of the seismic mass about the vertical axis through the c.g. is $I_z = M\rho_z^2 = 36 \times 10^3 \times 1.2^2$ kg·m².

Substituting these quantities into Eq. (7.8) gives

$$f_\gamma = \frac{1}{2\pi} \left[\frac{10.08 \times 10^6 \times 12(1.04 + 0.53)}{36 \times 10^3 \times 1.2^2} \right]^{1/2}$$

$$= 9.6 \text{ Hz}$$

Rotation about Horizontal Axis

As indicated earlier, an uncoupled rotational mode about Ox or Oy can occur only in a c.g.-type mounting. In practice a c.g. mounting is designed to have the isolator portions symmetrical in relation to the Ox and Oy axes, because there is no reason to do otherwise. With this symmetry these rotational modes are obviously uncoupled.

Therefore, we need not derive the conditions that an unsymmetrical layout of a c.g. mounting should satisfy in order to have uncoupled rotational modes about Ox and Oy. The reason for considering an unsymmetrical layout (Fig. 7.10) when discussing the coupling of horizontal translation with rotation about the vertical axis was that the result is applicable also to a base mounting.

Nevertheless, it may be of interest to remark that the conditions which must be satisfied by a c.g. mounting on a set of any number of identical isolators in an unsymmetrical layout, in order that its rotational modes about Ox and Oy should be uncoupled, turn out to be the same as those already established in other contexts [Eqs. (7.1) and (7.5)].

Natural Frequency about Horizontal Axis

The natural frequencies about the horizontal axes are calculated using equations of the same form as Eq. (7.7) but with angular stiffness and moment of inertia appropriate to the mode of oscillation; thus, the natural rotational frequencies about Ox and Oy are, respectively,

$$f_\alpha = \frac{1}{2\pi}\left(\frac{K_\alpha}{I_x}\right)^{1/2}, \quad f_\beta = \frac{1}{2\pi}\left(\frac{K_\beta}{I_y}\right)^{1/2}. \tag{7.9}$$

The values of K_α and K_β are derived in the following way. Referring to Fig. 7.12, if the seismic mass rotates through a small angle about a horizontal axis the vertical restoring force at each isolator is equal to the product of its vertical deflection and its vertical stiffness. Thus, the restoring force developed at each isolator by a small angular displacement α about Ox is $\alpha a_y k_z$. The restoring torque developed by each isolator is $\alpha a_y^2 k_z$. Therefore, the total restoring torque is $4\alpha a_y^2 k_z$ and the restoring torque per unit angular displacement α about the Ox axis is

$$K_\alpha = 4a_y^2 k_z \tag{7.10}$$

The corresponding expression for rotation about the Oy axis is

$$K_\beta = 4a_x^2 k_z \tag{7.11}$$

If in some unusual circumstances the isolators cannot be placed symmetrically, then obviously the sum of the squared values of the isolator coordinates is used instead of $4a_x^2$ and $4a_y^2$ in Eqs. (7.10) and (7.11).

Example 7.6. Calculate the natural frequencies of the seismic mounting in Example 7.4 and Fig. 7.12 in its uncoupled rotational modes about the horizontal axes, which can be regarded as principal axes of inertia. The total seismic mass is $M = 1.2$ tonnes (from Example 7.4) and the radii of gyration are estimated to be $\rho_x = 0.45$ m and $\rho_y = 0.35$ m.

From Example 7.4 the vertical static stiffness of one spring is 118 N/mm (118×10^3 N/m). For a helical spring this is taken as the vertical dynamic stiffness k_z.

The restoring torque of the four springs per unit displacement about the Ox axis is from Eq. (7.10)

$$K_\alpha = 4a_y^2 k_z = 4 \times 0.63^2 \times 118 \times 10^3 \text{ N·m/rad}$$

The moment of inertia about the Ox axis is

$$I_x = M\rho_x^2 = 1.2 \times 10^3 \times 0.45^2 \text{ kg} \cdot \text{m}^2$$

Substituting for K_α and I_x in Eq. (7.9) gives

$$f_\alpha = \frac{1}{2\pi} \left(\frac{4 \times 0.63^2 \times 118 \times 10^3}{1.2 \times 10^3 \times 0.45^2} \right)^{1/2} = 4.4 \text{ Hz}$$

Similarly, for the rotational mode about the Oy axis,

$$k_\beta = 4a_x^2 k_z = 4 \times 0.38^2 \times 118 \times 10^3 \text{ N} \cdot \text{m/rad}$$

$$I_y = M\rho_y^2 = 1.2 \times 10^3 \times 0.35^2 \text{ kg} \cdot \text{m}^2$$

Substituting these values in Eq. (7.9) gives

$$f_\beta = \frac{1}{2\pi} \left(\frac{4 \times 0.38^2 \times 118 \times 10^3}{1.2 \times 10^3 \times 0.35^2} \right)^{1/2} = 3.4 \text{ Hz}$$

COUPLED MODES OF A BASE-TYPE MOUNTING

In a base mounting on a set of identical isolators in a layout satisfying the conditions given by Eqs. (7.1) and (7.5), horizontal-x translation is coupled only with rotation about Oy (Fig. 7.8), so that the path of any point in the mass remains in a vertical plane parallel to the xz plane. Similarly, horizontal-y translation is coupled only with rotation about Ox so that the path of any point in the mass remains in a vertical plane parallel to the yz plane. For brevity we can refer to these motions as the coupled modes in the xz and yz planes, respectively.

It is not obvious that there are two modes of vibration in the xz plane and two in the yz plane. The nature of the two modes in each plane is perhaps best visualized in relation to their uncoupled counterparts in the c.g. mounting. The horizontal-x motion that is a pure translation in a c.g. mounting, is accompanied in a base mounting by rotation about Oy as indicated in Fig. 7.14a. This mode, in which the mass tilts clockwise as the c.g. moves to the right of its rest position, is the lower rocking mode (i.e., having the lower of the two natural frequencies in the xz plane). The mode that is an uncoupled rotation about Oy in a c.g. mounting, becomes in a base mounting the higher rocking mode, in which the mass tilts anticlockwise as the c.g. moves to the right (Fig. 7.14b). Likewise, there is a lower and a higher rocking mode in the yz plane. Our object is to calculate the natural frequencies in these coupled modes.

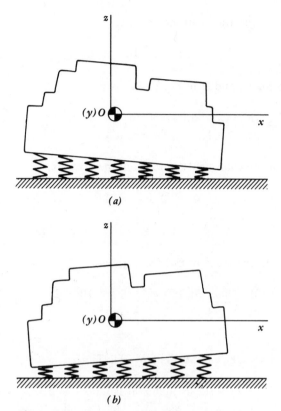

Figure 7.14.   Coupled modes of base-type mounting.

The method to be presented is an extension of a method introduced by Crede (1951) for calculating the coupled frequencies of a base mounting on four isolators having two planes of symmetry (Fig. 7.15). Therefore, we first recall Crede's method and then show how it can be applied to a base mounting on an unsymmetrical arrangement of any number of isolators.

The notation used in this book differs from that of Crede (1951), but is consistent with that of Crede and Ruzicka (1976). Referring to Fig. 7.15, the natural frequencies in the xz plane are denoted by $f_{x\beta}$, the suffix indicating that the modes concerned involve x-translation coupled with angular-β rotation about Oy. The natural frequencies in the yz plane involving y-translation coupled with angular-α rotation about Ox are denoted by $f_{y\alpha}$.

Four-Isolator Mounting with Two Planes of Symmetry

Methods of calculating the coupled natural frequencies of the system in Fig. 7.15 have been given by various investigators, perhaps the first being Hull (1937). The best known and most convenient method is Crede's, which involves the use of dimensionless ratios of the physical quantities concerned. Referring first to the coupled modes in the xz plane, the dimensionless ratios are

$\dfrac{f_{x\beta}}{f_z}$ = ratio of coupled mode frequency to uncoupled vertical natural frequency. The latter is evaluated using Eq. (7.2) (see Example 7.2).

$\dfrac{\rho_y}{a_x}$ = ratio of radius of gyration of seismic mass about Oy, to half-distance between isolators in the x-direction.

$\dfrac{a_z}{\rho_y}$ = ratio of height of c.g. above horizontal elastic plane of isolators, to radius of gyration about Oy. The elastic axis has been defined in the discussion relating to Fig. 7.9. For design purposes, when the height of the isolator is small compared with a_z the latter may be measured to the mid-height of the isolators.

$\dfrac{k_x}{k_z}$ = ratio of horizontal to vertical stiffness of an isolator.

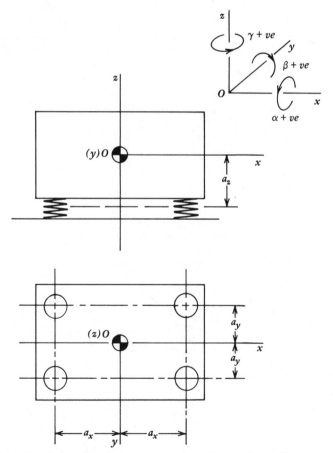

Figure 7.15.   Four-isolator base-type mounting with two planes of symmetry.

These dimensionless ratios are arranged in the two groups, which are plotted as ordinate and abscissa in Fig. 7.16. The coupled frequencies ($f_{x\beta}$) to be determined are contained in the ordinate $(\rho_y/a_x)(f_{x\beta}/f_z)$. Each curve in Fig. 7.16 shows the relationship between this ordinate and the abscissa $(\rho_y/a_x)(k_x/k_z)^{1/2}$ for a particular value of a_z/ρ_y.

The derivation of the mathematical relationships represented by the set of curves in Fig. 7.16 can be found in Crede (1951, pp. 53–55) and need not be

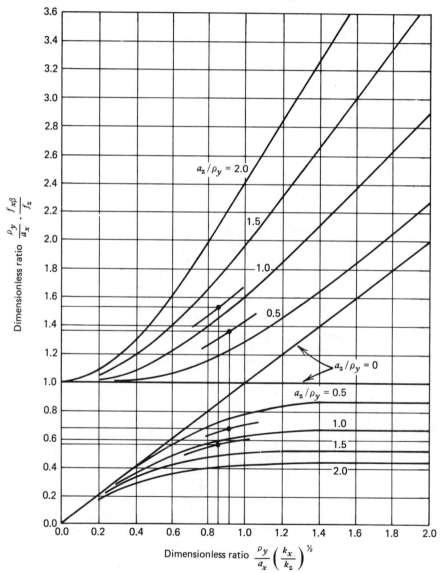

Figure 7.16. Design data for calculating coupled natural frequencies of base-type mounting (adapted from Crede 1951).

reproduced here. Fig. 7.16 shows only a few curves for reference in the present context: more comprehensive graphs can be found in Crede (1951, Fig. 2.8), and Crede and Ruzicka (1976, Fig. 30.18).

Crede's data are used in the following way to find the coupled natural frequencies of a four-isolator mounting (Fig. 7.15). For given values of a_x, a_z, ρ_y and k_x/k_z, calculate the value of the abscissa (Fig. 7.16) and calculate a_z/ρ_y. For the calculated value of the abscissa, read the ordinate to each of the two curves in Fig. 7.16 marked with the particular value of a_z/ρ_y. Extract from these the two values of $f_{x\beta}$ by making use of the known values of ρ_y/a_x and f_z. For the coupled frequencies in the yz plane use the same procedure but substitute $f_{y\alpha}$ for $f_{x\beta}$, ρ_x/a_y for ρ_y/a_x, and a_z/ρ_x for a_z/ρ_y. Example 7.7 shows how the method is applied to an unsymmetrical mounting on any number of isolators.

Unsymmetrical Mounting on Any Number of Isolators

The coupled natural frequencies of a base mounting on any number of isolators in any layout, symmetrical or otherwise, that satisfies the conditions defined by Eqs. (7.1) and (7.5) can be found by using Crede's data for a four-isolator mounting with two planes of symmetry. All that is necessary is to replace the distances a_x and a_y (Fig. 7.15) with the quantities X and Y, the root mean square values of the x and y coordinates of the isolator positions, and follow the procedure described in relation to Fig. 7.15 (see Example 7.7). The author's proof of the validity of the procedure is given elsewhere (Macinante, 1960).

In all the discussion of the base mounting in this book we assume that the isolators are on a common horizontal plane, as is normal in practice. It may be metioned that Crede's data can be used also to calculate the coupled natural frequencies of a base mounting on isolators which are not all on the same plane (Waldersee, 1961).

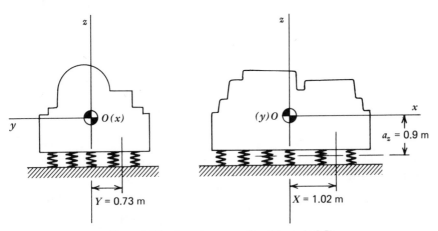

Figure 7.17.   Base-type mounting (Example 7.7).

Example 7.7. Calculate the coupled natural frequencies of the mounting in Fig. 7.17. The isolators are those defined in Examples 7.2 and 7.5. The isolator layout is that determined in Examples 7.1 and 7.4 to give an uncoupled vertical mode and no cross-coupling. The radii of gyration are estimated to be $\rho_x = 0.8$ m, and $\rho_y = 1.2$ m, and the height of the c.g. above the horizontal elastic plane of the isolators is $a_z = 0.9$ m.

In preparing to use Fig. 7.16 we collect the following data. In effect, we reduce the twelve-isolator unsymmetrical mounting to an equivalent four-isolator symmetrical mounting (Fig. 7.15).

f_z = 11.9 Hz, from Example 7.2

a_x = 1.02 m. This is derived as the root mean square value (X) of the x coordinates of the isolators. From Table 7.1, $X^2 = 12.46/12 = 1.04$ m^2; therefore, $X = 1.02$ m.

a_y = 0.73 m. This is the root mean square value of the y coordinates of the isolators. From Table 7.1, $Y^2 = 6.32/12 = 0.53$ m^2; therefore, $Y = 0.73$ m.

$\dfrac{k_x}{k_z}$ = 0.6 from Example 7.5.

The abscissa (Fig. 7.16) is

$$\frac{\rho_y}{a_x}\left(\frac{k_x}{k_z}\right)^{1/2} = \frac{1.2}{1.02}(0.6)^{1/2} = 0.91$$

and

$$\frac{a_z}{\rho_y} = \frac{0.9}{1.2} = 0.75$$

Reading from Fig. 7.16, the two values of the ordinate, corresponding to abscissa 0.91 and $a_z/\rho_y = 0.75$, are 0.68 and 1.35.

That is,

$$\frac{\rho_y}{a_x}\frac{f_{x\beta}}{f_z} = 0.68, 1.35$$

or

$$f_{x\beta} = \frac{a_x}{\rho_y}f_z(0.68, 1.35)$$

Substituting the values of a_x, ρ_y, and f_z, the natural frequencies in the coupled modes in the xz plane are

$$f_{x\beta} = \frac{1.02 \times 11.9 \times 0.68}{1.2} = 6.9 \text{ Hz}$$

and

$$f_{x\beta} = \frac{1.02 \times 11.9 \times 1.35}{1.2} = 13.7 \text{ Hz}$$

The coupled natural frequencies in the yz plane are found in a similar way. The abscissa in Fig. 7.16 is now

$$\frac{\rho_x}{a_y} \left(\frac{k_y}{k_z}\right)^{1/2} = \frac{0.8}{0.73} (0.6)^{1/2} = 0.85$$

and, assuming $k_x = k_y$,

$$\frac{a_z}{\rho_x} \cdot = \frac{0.9}{0.8} = 1.12$$

Reading from Fig. 7.16, the two values of the ordinate, corresponding to abscissa 0.85 and $a_z/\rho_x = 1.12$, are 0.56 and 1.53.

That is,

$$\frac{\rho_x}{a_y} \frac{f_{y\alpha}}{f_z} = (0.56, 1.53)$$

or

$$f_{y\alpha} = \frac{a_y}{\rho_x} f_z(0.56, 1.53)$$

Substituting the values of a_y, ρ_x, and f_z, the natural frequencies in the coupled modes in the yz plane are

$$f_{y\alpha} = \frac{0.73 \times 11.9 \times 0.56}{0.8} = 6.1 \text{ Hz}$$

and

$$f_{y\alpha} = \frac{0.73 \times 11.9 \times 1.53}{0.8} = 16.6 \text{ Hz}$$

REFERENCES

Crede, Charles E. (1951). *Vibration and Shock Isolation*, Wiley, New York.

Crede, Charles E., and Jerome E. Ruzicka (1976). Theory of Vibration Isolation, in *Shock and Vibration Handbook*, Chap. 30, Cyril M. Harris and Charles E. Crede, Eds., McGraw-Hill, New York.

Den Hartog, J. P. (1956). *Mechanical Vibrations*, 4th ed., McGraw-Hill, New York.

Hull, E. H. (1937). The Use of Rubber in Vibration Isolation, *Trans. Am. Soc. Mech. Eng.*, **59**, A109–A114.

Inglis, Charles Edward (1944). Mechanical Vibrations: Their Cause and Prevention, *J. Inst. Civil Eng.*, **22**, 312–357.

Macinante, J. A. (1960). Seismic Mountings for Large Machine Tools, *Engineer*, **210**, 880–883.

Waldersee, J. (1961). Natural Frequencies of Large Seismic Mountings on Non-coplanar Isolators, *Engineer*, **211**, 115.

8

Mountings for Machinery on Suspended Floors

Industrial plant and machinery is probably responsible for most vibration nuisance, and users of machinery that may create undue vibration should take the fullest precautions at the earliest possible stage. In many cases of vibration nuisance, the trouble could have been avoided had expert advice been obtained before the machinery was installed and the correct form of foundation or anti-vibration mounting provided. Cases have also been encountered where an isolating system has been provided without sufficient consideration of the fundamental principles of vibration isolation, the results being ineffective. . . .

R. J. STEFFENS (1966, p. 19)

Steffens' remarks about the need for early consideration of vibration control requirements, and for adequate attention to basic principles, are no less relevant today. In this book (Chapter 3) we emphasize the need for careful consideration of the vibration factor at an early stage in the design of a building, and indicate ways in which vibration trouble may be forestalled by design action relating to the source, the transmission path, and the receiver of the vibration. We also draw attention (Chapter 6) to what appears to be indifference to basic principles in the continued use of the one-mass design model in circumstances where it is clearly inadequate. The aim of the present chapter is to dispel this indifference in relation to the design of mountings for machinery on suspended floors.

If a mounting is to be designed for an existing machine, we assume that the decision to provide the mounting has been made only after considering and perhaps trying alternative lines of action discussed in Chapter 3: change in site of machine, modification of its operating conditions, and dynamic balancing.

If the mounting is for a proposed new installation we assume that in specifying the type of machine, its balance quality, and vibration severity rating, an effort has been made to minimize the vibration that will be generated by the machine and that, nevertheless, a seismic mounting is considered necessary to ensure that the vibration transmitted to the structure, to sensitive equipment, and/or to the occupants of adjacent areas will be within acceptable levels as discussed in Chapter 4.

FLOOR FLEXIBILITY DEMANDS TWO-MASS MODEL

The role of the design model and the considerations involved in the choice of model for source and receiver mountings are discussed in Chapter 6, where reasons are given for abandoning the one-mass model and adopting the two-mass, or even a more comprehensive model.

The practical application of the two-mass model in the design of mountings subjected to periodic, transient, and random vibration is indicated in general terms in Chapter 6. The present chapter deals specifically with mountings for sources of periodic vibration. For brevity we use the terms "machine" and "machinery" to include engines, turbines, motors, compressors, fans, blowers, generators, pumps, metal-working and other mechanical processing machines, vibrators, shakers, mixers, vibratory conveyors, and any other source of periodic vibration that may be found in office and residential buildings, and in factories.

The dominating reason for rejecting the one-mass model is that it does not involve the flexibility of the floor as a design factor. In the days when the heating, ventilating, and air-conditioning plant could be installed on basement or "on-grade" floors the one-mass model was applied with confidence, on the assumption that a floor formed directly on the ground could be regarded as rigid. However, even in these circumstances the assumption is of uncertain validity because of the compliance of the ground, and because the floor may become suspended after some time as a result of settlement of the ground under the floor.

Today, for reasons of economy of plant layout and because basement and ground-level space is required for vehicle parking and other purposes, plant must be installed at upper levels and hence on suspended floors. A suspended floor is designed to have a certain elastic deflection in service, and is therefore capable of vibrating. The greater the deflection of the floor under its own weight plus superposed load, the lower will be its natural frequency. Only an unusually stiff floor is likely to have its fundamental natural frequency high enough to justify neglect of the floor flexibility in the design of a machine installation on the floor.

Changes in design and construction practice in recent decades are resulting in a lowering of the stiffness and natural frequency of suspended floors. The reasons for this are well documented in two publications of the United States Bureau of Standards, on the subject of structural deflections. One is a broad survey of the literature and state-of-the-art in relation to structures in general (Galambos et al., 1973); the other (Crist and Shaver, 1976) deals in particular with floors. In the background to the survey the authors say (p. 2):

In design and construction in the past, deflections have been relatively small compared to member sizes or average building size, thus, they have not been a dominant concern for the structural designer. Within this decade the rapid increase of building construction costs have been pressing the industry for more economical methods of construction; new materials have become available which result in low effective material modulus and overall reduced structural stiffness; and more sophisticated and accurate methods of design and analysis are being used. For these reasons deflections have become more significant than in the past with respect to design control. . . .

Crist and Shaver discuss the dynamic factor in the design of floor systems. After a critical review of the development of traditional static stiffness criteria they discuss criteria from the vibration viewpoint, particularly in relation to human perception of floor vibration. In the background to this discussion the authors say (pp. 1–2):

Rigorous design procedures, higher allowable stresses and new construction materials result in effectively more flexible floor systems than have been constructed in the past. Traditionally, floor deflection considerations have been essentially ignored or "taken care of" by the present criterion which requires that a horizontal member shall not have a deflection greater than 1/360 of the span for a prescribed live load. More flexible floor systems are being introduced in the modern structure and they frequently show a lack of serviceability because of unsatisfactory deflection performance, i.e., vibration disturbance to occupants, rattling cupboards that set on the floor, vibration of ceiling covers (attached to floors), cracking, nonload bearing partition damage, etc. . . .

Floor design to avoid unpleasant vibration resulting from humans walking on the floor is one of the problems associated with the use of high strength materials, larger spans, and more flexible construction. We now discuss in detail another problem that is aggravated by the flexibility of floors in modern buildings, that of the design of mountings for machinery installed on such floors.

The need for consideration of the influence of floor stiffness in the design of mountings is indicated by comparing the permissible range of floor deflection with the normal range of deflection of seismic mountings. The traditional permissible deflection of 1/360 of the span, applied to floors with 10–20 m span, allows a deflection in the range 28–56 mm, which is well within the usual range of vertical static deflection of seismic mountings.

The need is indicated also by comparing natural frequencies. Many floors have a natural frequency in the range 10–30 Hz (e.g., Steffens 1966, p. 16). Prestressed concrete floors are reported to have natural frequencies as low as 4 Hz. (Caro et al., 1975, Table 1). This overall range of floor natural frequency (4–30 Hz) overlaps the normal range of vertical natural frequency of seismic mountings (2–20 Hz).

Since the range of vibration frequency generated by machinery overlaps these ranges, the excitation frequency may coincide with the natural frequency of the floor, or that of the mounting, or both, thereby causing increased transmission of vibration. Obviously, the dynamic characteristics of the floor should be considered as a factor in the design of the mounting, and this calls for at least a two-mass model.

ONE-MASS MODEL AND "OVERKILL"

Although it has been acknowledged for nearly two decades (e.g., Eberhardt, 1966) that floor flexibility should be taken into account in the design of machine installations, the one-mass model has not been abandoned. It has been given a new lease of life in a practice that has been referred to as "overkill"—a mounting for a machine on a suspended floor is designed to have a much greater static deflection than would be considered necessary for the same machine installed on a "rigid" floor. For example, Eberhardt (1966, p. 59), and U.S. General Services Administration (1971, Table 1) tabulate suggested values of isolator deflection that are intended to compensate for floor deflection for various kinds of machinery and for several values of the floor span.

Similarly, the American Society of Heating, Refrigerating and Air-Conditioning Engineers (ASHRAE, 1980, Systems Handbook, p. 35.23) states that accepted practice is to increase isolator deflection as floor span and potential floor deflection increase. The Handbook (Table 27, p. 35.21) recommends isolator deflections that are based on the experience of acoustical and mechanical consultants, and vibration control manufacturers and installers. However, in presenting these recommendations the Handbook (p. 35.23) cautions that isolator deflection is not the sole controlling variable, and gives some guidance on the influence of the floor characteristics, based on the theory of the two-mass model. The nature of this guidance is indicated in the discussion that follows on the theory of the two-mass model.

The procedure presented in this chapter involves design variables that are more meaningful and specific than those used in the overkill approach: mounting natural frequency rather than mounting deflection; floor natural frequency rather than span. In addition, the procedure is more comprehensive in that it involves the damping of the mounting and that of the floor, and the mass ratio of the installation.

THEORETICAL BASIS

The acceptability of design practice based on experience is greatly enhanced if the practice is seen to be supported by theory. Therefore, let us see if vibration theory can show how the floor characteristics influence vibration transmission. Specifically, we wish to determine how the magnitude of vibration transmitted into the structure supporting the floor is influenced by the characteristics of the floor and those of the mounting. The results will provide a means of determining, for a machine that is to be installed on a suspended floor, what should be the characteristics of the mounting to ensure minimum transmission of vibration.

As far as the author is aware, the first application of the theory of the two-mass system to the design of a machine installation on a suspended floor is that described by Eason (1919; 1923, Appendix II). Eason derived an expression for the natural frequencies of free vibration of an undamped two-mass sysem representing the installation, so that the hardware could be designed to have its natural frequencies well separated from the excitation frequency, and hence operate well away from a resonance condition. Eason's derivation is based on that of Stodola (1905, pp. 355–358) in relation to the vibration of turbine foundations.

While the avoidance of resonance is essential for vibration control in that it eliminates the possibility of magnification of the vibration, it does not necessarily result in attenuation. The converse is true, however, that a two-mass system designed to attenuate, that is, to have a low ratio of transmitted to applied amplitude, cannot be operating at or near a resonance of the system. Therefore, our objective is to determine how the transmissibility ratio is related to the design variables.

The theory of the forced vibration of two-mass systems is well documented (e.g., Den Hartog, 1956, Chap. 3; Snowdon, 1968, Chaps. 3 and 4). The details of expressions for the transmissibility ratio depend, of course, on the way transmissibility is defined, and this in turn depends on the intended application of the results. For example, the theory may be applied to the design of vibration absorbers, containers for the transport of delicate equipment in vehicles, and mountings for machinery and sensitive equipment in buildings.

With reference to the design of machine mountings the ASHRAE Handbook (1980 Systems, pp. 35.22 to 35.23) refers to the one degree-of-freedom (d-f) system as

> an ideal situation which almost never exists in buildings, because all equipment room floors deflect under load except for those at grade locations or in basements where the concrete remains in intimate contact with soil. . . .

The Handbook mentions the difficulty of analysis of a multiple d-f system that would be necessary to represent conditions in an actual building, and states that for purposes of simplification this system can be visualized as an undamped two d-f system.

For an undamped one d-f system the transmissibility is determined only by the ratio of the forcing frequency to the natural frequency of the mounting. For an undamped two d-f system the transmissibility expression given in the Handbook involves, in addition, two design variables involving floor characteristics. These are the ratio of isolator stiffness to floor stiffness, and the ratio of forcing frequency to floor natural frequency. The Handbook gives some general guidance on the application of these factors in practical design situations. In particular, the Handbook (p. 35.23) warns that excessive vibration transmission as well as vibration of both equipment and floor can result if the forcing frequency is equal or close to the natural frequency of the floor.

The theory of the undamped system is a useful guide in designing to avoid resonance, but may give a misleading indication of the transmissibility ratio and the response amplitudes of a system having even the moderate damping (damping ratio less than 0.2) that is typical of actual installations. We now describe the theoretical basis of a method developed by Macinante and Simmons (1976, 1977) which takes damping into account.

Design Model

The design model is that given in Fig. 6.6b, which, for convenience of reference, is reproduced in Fig. 8.1c with added symbols which are defined below. For ease of identification we use the subscript "m" for mounting, and "f" for floor.

m_m seismic mass comprising machine, plus inertia block, if any.

m_f equivalent mass of floor; that is, the mass in a single degree of freedom system having the same vertical natural frequency f_f and stiffness k_f as the floor alone (i.e., not loaded by the installation).

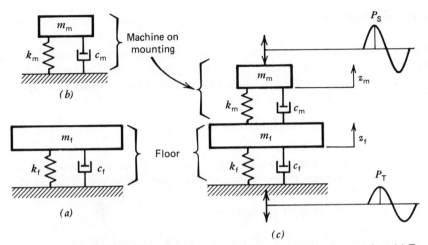

Figure 8.1. Subsystems representing (*a*) floor and (*b*) machine on seismic mounting. (*c*) Two-mass model of machine installation on suspended floor.

k_m vertical stiffness of the mounting; that is, total stiffness of all the isolators. For isolators having nonlinear load/deflection properties the stiffness is the dynamic stiffness appropriate to small displacements about the rest position, as discussed in relation to Fig. 5.8.

k_f stiffness of the floor; that is, vertical force per unit deflection of floor where mounting is to be installed.

c_m equivalent viscous damping coefficient of the mounting; that is, that of the isolators and/or separate damping elements.

c_f equivalent viscous damping coefficient of the floor.

The alternating force that is generated by the machine is assumed to be the sinusoidal force

$$P_S(t) \ = \ P_S \sin \omega_s t$$

where P_S = force amplitude
ω_s = $2\pi f_s$ = angular frequency (rad/s)
f_s = cyclical frequency (Hz)
t = time

This force is assumed to act in a vertical line passing through the center of gravity of mass m_m and of m_f. It is assumed that both masses vibrate only vertically. Because of this limitation, the behavior of the installation in modes other than the vertical must be considered independently of the design data presented below, which relate only to vertical vibration (see Practical Application of the Design Data, Check Vibration Amplitudes).

Transmissibility Ratio

As the function of the mounting is to minimize force transmission into the structure supporting the floor, the transmissibility ratio, denoted by T_F, is defined as the ratio of the amplitude of the transmitted to that of the applied force.

$$\text{Transmissibility ratio } T_F = \frac{P_T}{P_S} \qquad (8.1)$$

where P_T = amplitude of transmitted force, that is, resultant of elastic and damping forces transmitted through elements k_f and c_f (Fig. 8.1c)

P_S = amplitude of applied force

The subscript "F" denoting "force" is used to distinguish the force transmissibility ratio just defined from the relative displacement form of transmissibility ratio used in Chapter 9.

The transmitted force is visualized as the total force acting around the perimeter where the floor joins the structure. In general the transmitted force is not in phase with the applied force, but the phase difference is ignored as it is of no practical importance in the design of the mounting.

The expression derived for T_F does not directly use the quantities just defined, but incorporates them in the following dimensionless ratios, which are more meaningful, and permit the results to be presented in a form that is more convenient for design purposes.

$$
\begin{aligned}
R_m &= m_m/m_f & &\text{mounting mass ratio} \\
\zeta_m &= c_m/c_{cm} & &\text{mounting damping ratio} \\
\zeta_f &= c_f/c_{cf} & &\text{floor damping ratio} \\
\rho_m &= f_m/f_s & &\text{mounting frequency ratio} \\
\rho_f &= f_f/f_s & &\text{floor frequency ratio}
\end{aligned}
$$

where c_{cm} = critical damping coefficient of mounting

c_{cf} = critical damping coefficient of floor

f_m = undamped natural frequency of the mounting in its uncoupled vertical mode, that is, the mounting system alone, assuming it is on a rigid support (Fig. 8.1b)

f_f = undamped natural frequency of the floor in its fundamental flexural mode, that is, the floor system alone, unloaded by the installation (Fig. 8.1a)

f_s = source vibration frequency

The natural frequencies f_m and f_s of the one-mass systems just defined are not the same as the two natural frequencies of the system formed by joining together the two one-mass systems. This is discussed later in relation to the resonance frequencies of the installation.

Three of the ratios, the mass ratio and the two damping ratios, are determined by the hardware of the machine, the mounting, and the floor. The two frequency ratios involve, in addition, the operating frequency of the machine.

The expression for T_F in terms of these design variables is lengthy and is not reproduced here because the numerical data calculated using the expression, and presented pictorially and graphically in the next section, make direct calculation from the expression unnecessary. The expression and its derivation are published elsewhere (Macinante and Simmons, 1977, Appendix A). However, it is desirable to give an outline of the derivation of the expression for T_F not only to establish its parentage, but also to show how, in the course of the derivation, expressions are found which can be used to check the vibration amplitudes of machine and floor. In addition, these expressions serve to illustrate, in terms of high displacement amplitudes of machine and floor, the phenomenon of resonance, which is otherwise demonstrated in terms of high force amplitude transmission.

Basis of Computations

In order to arrive at an expression for the ratio of transmitted to applied force, we must determine the transmitted force P_T. This is the resultant of the forces in the spring k_f and the damper c_f, which in turn depend on the vibratory motion of the floor. Since the motion of the floor is influenced by that of the mounting which is coupled to it, we must determine the motion of the coupled two-mass system.

We begin by applying Newton's second law of motion. Newton's laws and their applications are expounded in textbooks with varying degrees of rigor, ranging from the mathematically pure to the bland and unashamed use of the rule "force equals mass times acceleration." The latter can give valid results if applied with appreciation of the physical nature of the forces involved and with consistent use of a sign convention defining the positive directions of force and displacement (e.g., see Den Hartog, 1956, pp. 25, 80).

Referring now to the two-mass system in Fig. 8.1c, we apply this rule to each mass in turn. The forces acting on the mass m_m at any instant are the instantaneous values of (1) the applied or exernal force $P_S \sin \omega_s t$, (2) the spring force, which is the product of the stiffness k_m and the change in length of the spring $(z_f - z_m)$, and (3) the damping force, which is the product of the damping coefficient c_m and the relative velocity across the damper $(\dot{z}_f - \dot{z}_m)$. Here we are using the "dot" notation mentioned in Chapter 2, Sinusoidal Vibration: the single dot denotes velocity, the first time derivative of displacement, and the double dot denotes acceleration, the second time derivative. The equation of motion of m_m is formed by equating the algebraic sum of these forces to $m_m \ddot{z}_m$, the product of the mass and its acceleration.

Similarly, the equation of motion of m_f is obtained by equating $m_f\ddot{z}_f$ to the algebraic sum of the instantaneous values of the forces acting on m_f. These are the spring forces acting above and below m_f, namely, $k_m(z_f - z_m)$ and $k_f z_f$; and the damping forces $c_m(\dot{z}_f - \dot{z}_m)$ and $c_f \dot{z}_f$.

The equation of motion of each mass involves the displacement and velocity of the other mass as well as its own. This mathematical coupling of the two equations reflects the physical coupling of the two masses, and requires their algebraic solution as simultaneous equations. Before solving the equations we replace the masses, stiffnesses, and damping coefficients with the dimensionless ratios defined earlier. At this stage we need to define a dimensionless displacement amplitude in order to sustain the dimensionless form of the design variables. We do this by expressing the displacement amplitude as a ratio of a static deflection δ_P which is defined (after Den Hartog, 1956, p. 88) as the deflection that the mounting would have if subjected to a static force equal to the amplitude P_S of the applied sinusoidal force:

$$\delta_P = \frac{P_S}{k_m} \tag{8.2}$$

This is an artificial kind of static deflection, not to be confused with the static deflection of the seismic mass under its own weight, defined in the discussion relating to Fig. 7.5. Thus, the displacement amplitudes of m_m and m_f, in dimensionless form, are Z_m/δ_P and Z_f/δ_P, respectively.

We now solve the equations in the conventional way for the steady-state response of the system, by assuming that both masses vibrate with sinusoidal motion at the frequency f_s of the applied force. In making this assumption we exclude from the solution the "starting transients," which occur when the excitation begins but decay after a short time leaving only the steady-state response. The masses vibrate with different amplitudes and different phase. The amplitudes are of practical importance; the phase is not. Solution of the equations yields expressions for the two amplitude ratios. We refer to these as the "general" expressions, in the sense that they involve the damping of both the mounting and the floor (see Macinante and Simmons, 1977, Appendix A, Eqs. A5 and A6).

The general expressions are not reproduced here, not merely because they are lengthy but also because they are formed from "packages" of mathematical terms grouped for convenience of algebraic manipulation in their derivation, and retained for use in the computer. Because of this, inspection of the expressions conveys nothing about the physical behavior of the system. Moreover, as will be shown later, one of the amplitude ratios—that of the floor—can be calculated without recourse to the general expression.

Having determined the amplitude of sinusoidal displacement of the floor, we can now form an expression for the amplitude P_T of the force transmitted to the structure that supports the floor. This force is the resultant of the alternating forces of amplitude $k_f z_f$ and $c_f \dot{z}_f$ in the spring k_f and damper c_f, respectively, which represent, in the design model, the elasticity and damping of the floor. By using the expression derived for the floor displacement, and its derivative velocity, and taking into account the phase relationship of the spring and damping forces, we find the required expression for P_T.

Finally, by expressing this force amplitude P_T as a ratio of the applied force amplitude P_S, then making some algebraic manipulations to sustain the dimensionless form of the variables, and denoting the frequency ratios f_m/f_s and f_f/f_s by the symbols ρ_m and ρ_f, respectively, to make the algebraic expressions more compact, we find the transmissibility ratio:

$$T_F = \frac{Z_f}{\delta_P} \frac{\rho_f}{R_m \rho_m^2} (\rho_f^2 + 4\zeta_f^2)^{1/2} \tag{8.3}$$

It is the first factor, the floor amplitude ratio, which, when written in terms of the system parameters including the damping ratios of both mounting and floor, makes the expression for T_F rather lengthy and cumbersome (see Macinante and Simmons, 1977, Appendix A, Eqs. A6 and A9).

For a completely undamped system the amplitude ratios and the transmissibility ratio reduce to the relatively simple forms: Floor amplitude ratio:

$$\frac{Z_f}{\delta_P} = \frac{R_m \rho_m^4}{D} \tag{8.4}$$

Mounting amplitude ratio:

$$\frac{Z_m}{\delta_P} = \frac{R_m \rho_m^4 + \rho_m^2 \rho_f^2 - \rho_m^2}{D} \tag{8.5}$$

Transmissibility ratio:

$$T_F = \frac{\rho_m^2 \rho_f^2}{D} \tag{8.6}$$

where the denominator D is

$$D = 1 - (R_m + 1)\rho_m^2 - \rho_f^2 + \rho_m^2 \rho_f^2 \tag{8.7}$$

Resonance Frequencies

By designing for low transmissibility, the requirement of avoiding resonance is automatically satisfied because resonance produces a high transmissibility. Nevertheless, the designer may wish to check the margin separating operating from resonance conditions, particularly if there is a possibility that a change in the operating conditions may bring the excitation frequency closer to a resonance frequency of the system. It will be seen in the data presented later that a small change in excitation frequency in the vicinity of a resonance frequency of a lightly damped system results in a large change in the transmissibility.

As noted earlier when defining the symbols, the natural frequencies f_m and f_f of the separate one-mass systems in Fig. 8.1 are not the same as the natural frequencies of the system formed by joining together the two one-mass systems.

Expressions for calculating the natural frequencies of the two-mass system can be derived by writing the equations of motion of the two masses in the manner described earlier in relation to the derivation of the expression for the transmissibility ratio, but now omitting the external force and the damping forces. The damping is ignored because the relatively low levels of damping in floors and mountings currently used have little influence on the natural frequencies. The equations so formed describe the free motion of the system after it has been set in motion by some initial or transient excitation.

The equations are solved by making the assumption that the free vibration of the system involves sinusoidal motion of both masses at a common natural frequency. After making substitutions to express the equations in terms of the mass ratio R_m and the natural frequencies f_m and f_f of the two subsystems and rearranging, we obtain the "frequency equation":

$$f_n^4 - (R_m f_m^2 + f_m^2 + f_f^2)f_n^2 + f_m^2 f_f^2 = 0 \qquad (8.8)$$

The solution of this quadratic equation in f_n^2 is

$$f_n^2 = \tfrac{1}{2}B \pm \tfrac{1}{2}(B^2 - 4C)^{1/2}$$

where

$$B = R_m f_m^2 + f_m^2 + f_f^2$$
$$C = f_m^2 f_f^2$$

Therefore

$$f_n = [\tfrac{1}{2}B \pm \tfrac{1}{2}(B^2 - 4C)^{1/2}]^{1/2} \qquad (8.9)$$

This yields the values f_{n1} and f_{n2} of the natural frequencies of the system in its two natural modes. In one mode the masses move in phase, the other antiphase.

We know intuitively that resonance occurs when the excitation frequency equals one or other of the natural frequencies of the system. This fact is embodied in the expressions [Eqs. (8.4)–(8.6)] for the amplitudes of the masses and the transmissibility ratio. The system is in resonance when the denominator [Eq. (8.7)] is zero, that is, when

$$1 - (R_m + 1)\rho_m^2 - \rho_f^2 + \rho_m^2 \rho_f^2 = 0 \qquad (8.10)$$

This may be rewritten in the following form by replacing ρ_m and ρ_f with f_m/f_s and f_f/f_s, respectively, and simplifying:

$$f_s^4 - (R_m f_m^2 + f_m^2 + f_f^2)f_s^2 + f_m^2 f_f^2 = 0$$

which is identical with Eq. (8.8) when $f_s = f_n$.

This means that the denominator is zero and the system is in resonance when $f_s = f_{n1}$ or $f_s = f_{n2}$. In theory the transmissibility ratio and the two amplitude ratios increase toward infinity. It is significant that the expressions for the two amplitude ratios and the transmissibility ratio all have the same denominator. When the denominator is zero the amplitude of each mass tends to become infinite—it is unthinkable, of course, that the amplitude of one of the masses could approach infinity without the other doing so—and the theoretically infinite displacement of the floor mass is consistent with the transmission of an infinite force P_T. In an actual system, of course, the responses are limited by the damping.

The relationship in Eq. (8.10) is illustrated in Fig. 8.2 for three values of the mass ratio $R_m = 0.1, 0.5, 1.0$. The undamped systems with these values of the mass ratio are defined later as Cases 1, 2, 3, respectively. There is a pair of curves for each value of the mass ratio. A point on one of these curves represents resonance at one of the natural frequencies ($f_s = f_{n1}$); a point on the other curve represents resonance at the other natural frequency ($f_s = f_{n2}$), as shown in the following example.

Example 8.1.   A machine on a seismic mounting having a vertical natural frequency 14 Hz is installed on a floor whose fundamental natural frequency is 18 Hz. The mass ratio is 0.1. At what excitation frequencies will resonance occur?

Referring to Eq. (8.9) and substituting $f_m = 14$, $f_f = 18$, and $R_m = 0.1$:

$$B = 0.1(14)^2 + (14)^2 + (18)^2 = 539.6,$$
$$C = (14)^2(18)^2 = 63504,$$

and the two natural frequencies of the system are found to be:

$$f_{n1} = 13.2 \text{ Hz} \qquad f_{n2} = 19.1 \text{ Hz}$$

A resonance occurs when the excitation frequency f_s is equal to one or other of these natural frequencies. In terms of the frequency ratios:

If $f_s = f_{n1} = 13.2$

$$\frac{f_m}{f_s} = \frac{14}{13.2} = 1.06$$

$$\frac{f_f}{f_s} = \frac{18}{13.2} = 1.36.$$

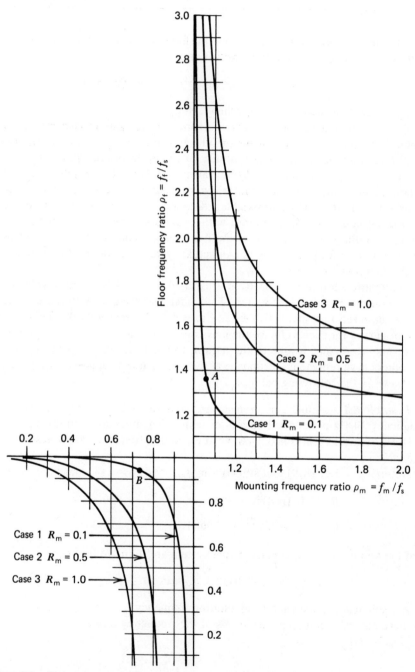

Figure 8.2. Resonance curves for undamped two-mass system with three values of mass ratio (for case numbers see Table 8.1).

This resonance is indicated by the point A in Fig. 8.2.
   If $f_s = f_{n2} = 19.1$

$$\frac{f_m}{f_s} = \frac{14}{19.1} = 0.73$$

$$\frac{f_f}{f_s} = \frac{18}{19.1} = 0.94.$$

This resonance is indicated by the point B in Fig. 8.2.

Vibration Amplitudes

The displacement amplitude ratios of mounting and floor can be calculated by using the general expressions referred to in the foregoing outline of the derivation of the expression for the transmissibility ratio [Eq. (8.3)]. The floor amplitude ratio is calculated more easily by using the fact that this ratio is the factor Z_f/δ_p in Eq. (8.3). Thus, if a system is designed to have a nominated value of T_F, the floor amplitude ratio is evaluated by substituting in Eq. (8.3) the values of T_F and the other relevant quantities, as illustrated later (Example 8.3).

   For a lightly damped system designed for operation well away from resonance, the amplitude ratios may be checked by using Eqs. (8.4) and (8.5) for an undamped system. This involves the assumption that the damping has a significant effect only in the vicinity of resonance. An example of the influence of damping on the resonance response is included in Example 8.3.

   The foregoing discussion is concerned only with the vertical vibration amplitudes. In a practical application the vertical vibration amplitude at any point on the seismic mass may be greater or less than the calculated value as a result

Table 8.1.   Case Numbers for Machine Installations

Mounting Damping Ratio ζ_m	Mass Ratio R_m	Floor Damping Ratio ζ_f		
		0	0.1	0.2
0	0.1	Case 1	Case 4	Case 7
	0.5	2	5	8
	1.0	3	6	9
0.1	0.1	Case 10	Case 13	Case 16
	0.5	11	14	17
	1.0	12	15	18
0.2	0.1	Case 19	Case 22	Case 25
	0.5	20	23	26
	1.0	21	24	27

of coupling of the vertical mode with a rotational mode about a horizontal axis (Fig. 7.1). At a point remote from the c.g. the resultant vertical component of the amplitude may be significantly greater than that calculated, because the component contributed by motion in a rotational mode about a horizontal axis increases with distance from the c.g. The total or resultant amplitude will be still greater, if there are other components of vibration at the point.

DESIGN DATA BASED ON TWO-MASS MODEL

With the object of providing data of immediate value in the design office, we have computed the transmissibility ratio T_F for values of the design variables in ranges covering the needs of normal practice. The results are presented as T_F surfaces and contours (three- and two-dimensional computer plots), which show how T_F is related to the design variables.

Three values (0.1, 0.5, 1.0) were nominated for the mass ratio, and three values (0, 0.1, 0.2) for each of the two damping ratios. These form 27 combinations which, for convenience of reference, are identified by the "case numbers" given in Table 8.1. For each of these combinations of mounting damping ratio, floor damping ratio and mass ratio, the transmissibility ratio was computed for the range of values of mounting frequency ratio and floor frequency ratio indicated below.

Transmissibility Surfaces

The way in which T_F is influenced by the two frequency ratios for any particular case is most clearly shown as a surface with T_F plotted as the height, and mounting and floor frequency ratios as the horizontal coordinates. Figure 8.3 shows the computer-drawn surfaces for nine representative cases, the first in each group of three in Table 8.1 (i.e., Cases 1, 4, 7, 10, . . .). The discontinuities in the ridges for Case 1 where T_F is changing rapidly are a consequence of the finite step size used in the computer when plotting the surfaces. Each surface was plotted with T_F computed for all combinations of the following values of the frequency ratios:

$$\rho_m = f_m/f_s: \quad 0.05, 0.10, 0.15, \ldots, 2.0$$

$$\rho_f = f_f/f_s: \quad 0.50, 0.55, 0.60, \ldots, 3.0$$

The symbols ρ_m and ρ_f are used mainly to simplify mathematical expressions involving the frequency ratios [e.g., Eqs. (8.3) through (8.7)]. In much of the discussion to follow, the ratios are expressed in terms of the two frequencies involved so that the significance of the individual frequencies will not be obscured by the notation.

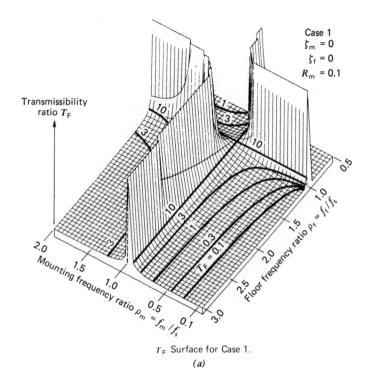

T_F Surface for Case 1.

(a)

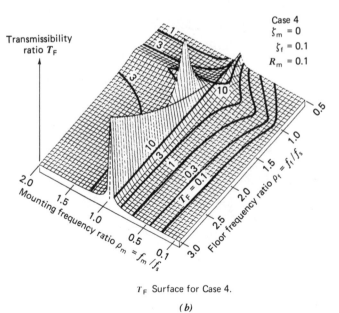

T_F Surface for Case 4.

(b)

Figure 8.3. Transmissibility-ratio surfaces for two-mass system with mass ratio 0.1, for three values of the damping ratio of the floor, and three values of the damping ratio of the mounting (for case numbers see Table 8.1). (Macinante and Simmons, 1981).

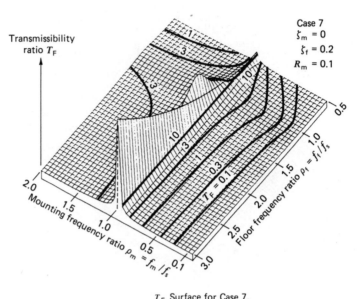

T_F Surface for Case 7.

(c)

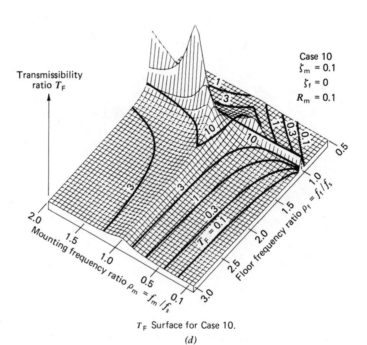

T_F Surface for Case 10.

(d)

Figure 8.3 (*continued*).

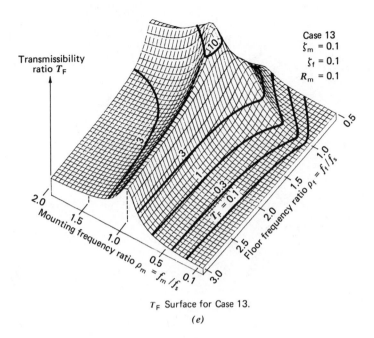

Case 13
$\zeta_m = 0.1$
$\zeta_f = 0.1$
$R_m = 0.1$

T_F Surface for Case 13.

(e)

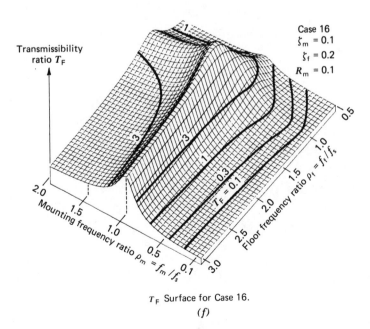

Case 16
$\zeta_m = 0.1$
$\zeta_f = 0.2$
$R_m = 0.1$

T_F Surface for Case 16.

(f)

Figure 8.3 (*continued*).

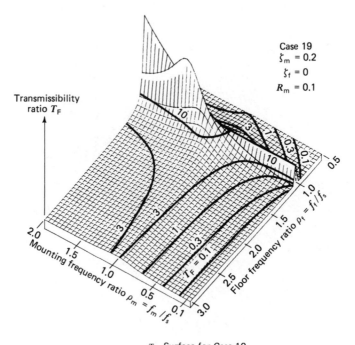

Case 19
$\zeta_m = 0.2$
$\zeta_f = 0$
$R_m = 0.1$

Transmissibility ratio T_F

Mounting frequency ratio $\rho_m = f_m/f_s$

Floor frequency ratio $\rho_f = f_f/f_s$

T_F Surface for Case 19.

(g)

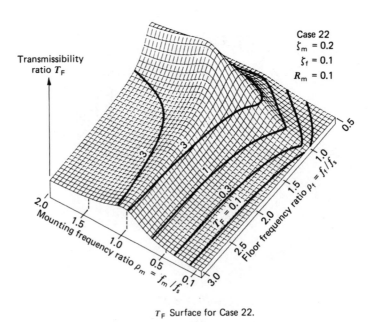

Case 22
$\zeta_m = 0.2$
$\zeta_f = 0.1$
$R_m = 0.1$

Transmissibility ratio T_F

Mounting frequency ratio $\rho_m = f_m/f_s$

Floor frequency ratio $\rho_f = f_f/f_s$

T_F Surface for Case 22.

(h)

Figure 8.3 (*continued*).

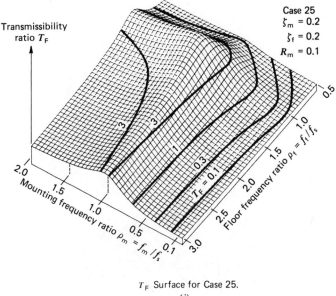

Case 25
$\zeta_m = 0.2$
$\zeta_f = 0.2$
$R_m = 0.1$

T_F Surface for Case 25.
(i)

Figure 8.3 *(continued)*.

Transmissibility Contours

Some contours of constant T_F were hand-plotted on the surfaces. More detailed contours were computer-plotted for five values of T_F (0.01, 0.03, 0.10, 0.30, 1.0) for each of the 27 cases in Table 8.1. Each set of contours was plotted with T_F calculated for all combinations of the following values of the two frequency ratios, and for additional values where necessary to fill in detail in the region of resonances where the contours change rapidly:

$$f_m/f_s: \quad 0.02, 0.04, 0.06, \ldots, 1.0$$

The upper limit satisfies practical needs, because the mounting natural frequency is normally—and of necessity, as illustrated later—made less than the excitation frequency

$$f_f/f_s: \quad 0.10, 0.12, 0.14, \ldots, 1.50$$

The earlier investigation (Macinante and Simmons, 1976) showed that there is little change in T_F with $f_f/f_s > 1.5$.

Figure 8.4 shows the orientation of a set of contours in relation to the corresponding surface. The surface is shown with the region of low T_F in the foreground, for this is the area of most interest to the designer of a mounting, while the contours are drawn in the normal way for two-dimensional graphs.

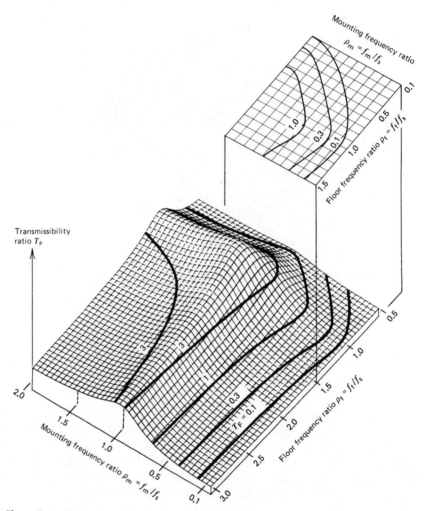

Figure 8.4. Orientation of transmissibility-ratio contour plots in relation to surface plots.

It is possible to present three sets of contours in a single diagram, without a confusion of overlapping contours, by grouping cases having a common mass ratio. For example, Fig. 8.5 shows the group of the sets of contours relating to Cases 2, 5, and 8. Two other groups are illustrated for reference in the discussion and the numerical example to follow. These are Fig. 8.6 for Cases 11, 14, and 17; and Fig. 8.7 for Cases 20, 23, and 26. The contours for all cases listed in Table 8.1 are given in the Appendix.

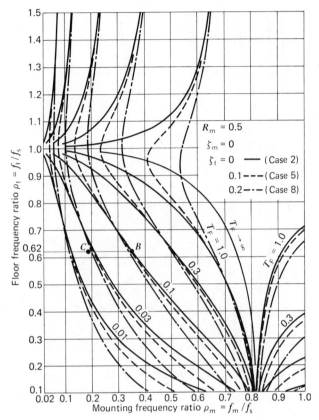

Figure 8.5. Transmissibility-ratio contours for two-mass system with mass ratio 0.5, and undamped mounting (for case numbers see Table 8.1). (Macinante and Simmons, 1981).

SUMMARY OF RESPONSE OF TWO-MASS MODEL

The following general qualitative conclusions, drawn from inspection of the T_F surfaces and contours, show how T_F is influenced by the mounting and floor frequency ratios and damping. They provide a basis for selecting combinations of frequency ratios that yield low T_F and, incidentally, they show that the practice of designing for an exaggerated static deflection (overkill) involves the risk that high T_F may result from an unlucky coincidence of floor resonance frequency and excitation frequency.

Resonance Conditions

For the completely undamped system (Cases 1, 2, 3) the T_F surface has two ridges (e.g., Fig. 8.3, Case 1) along which T_F is very high, theoretically

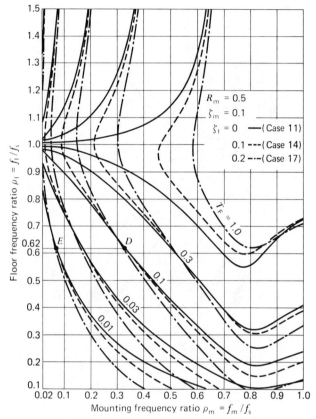

Figure 8.6.   Transmissibility-ratio contours for two-mass system with mass ratio 0.5, and mounting damping ratio 0.1 (for case numbers see Table 8.1). (Macinante and Simmons, 1981).

tending to infinity. A point on a ridge represents a resonance that can occur if the excitation frequency coincides with one or other natural frequency of the system, as discussed in relation to Fig. 8.2 and illustrated in Example 8.1. The curves in Fig. 8.2 for $R_m = 0.1$ are, in effect, a plan view of the ridges shown in Fig. 8.3, Case 1.

In areas near these ridges obviously a small change in frequency ratio can cause a large change in T_F. The rapidity of the rate of change of T_F with frequency ratio is illustrated for Case 2 in Fig. 8.5 in which the curve representing the theoretically infinite T_F for Case 2 is close to the $T_F = 1.0$ contour.

With an undamped mounting on a damped floor (Cases 4–9) there is a ridge of high T_F which can be attributed to resonance of the mounting. This can be seen in Fig. 8.3 for Cases 4 and 7. In the contour diagrams in Fig. 8.5, Cases

5 and 8, the ridge lies between the two $T_F = 1$ contours at the lower right of the figure.

With a damped mounting on an undamped floor (Cases 10–12, 19–21) there is a ridge of high T_F which can be attributed to floor resonance as illustrated in Fig. 8.3, Cases 10 and 19. In the contour diagrams in Fig. 8.6, Case 11, and Fig. 8.7, Case 20, the ridge is in the area between the two $T_F = 1$ contours which converge in the vicinity of $f_f/f_s = 1$.

With both mounting and floor damped (Cases 13–18, 22–27), a region of high T_F occurs where both frequency ratios are close to unity, as illustrated in Fig. 8.3 for Cases 13, 16, 22, and 25. The overall effect of system damping is seen by noting that the T_F surfaces become lower and more rounded with increase in damping of the mounting and with increase in damping of the floor.

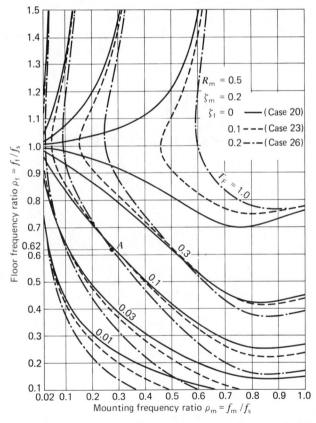

Figure 8.7. Transmissibility-ratio contours for two-mass system with mass ratio 0.5, and mounting damping ratio 0.2 (for case numbers see Table 8.1). (Macinante and Simmons, 1981).

Stiff Floor

If the floor natural frequency is much higher than the excitation frequency ($f_f/f_s > 1.5$) the behavior of the system is described by the part of the T_F surface identified in Fig. 8.8 as the "stiff floor" zone.

In the part of this zone where the mounting natural frequency is less than the excitation frequency ($f_m/f_s < 1$), the T_F contours are roughly parallel to the f_f/f_s axis; hence, the floor frequency ratio has little influence on T_F. This is seen in the nine examples in Fig. 8.3 and is illustrated in Fig. 8.9 for all 27 cases for the particular value $T_F = 0.1$.

In the stiff floor zone it is evident, from the shape of the "end wall" where $f_f/f_s = 3$ (Figs. 8.3 and 8.8), that T_F is determined primarily by the mounting frequency ratio. The response is like that of a one-mass system comprising a spring-mounted mass on a rigid floor. The resonance frequency varies slightly with the mass ratio. This is indicated in Fig. 8.2 for an undamped system by the separation of the curves at the top of the illustration.

Since resonance occurs for values of f_m/f_s a little greater than unity, a simple and conservative basis for the design of a mounting on a floor having $f_f/f_s >$

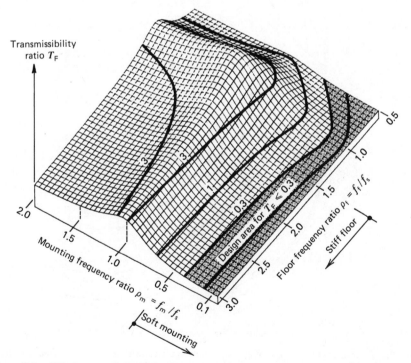

Figure 8.8. "Design area," "stiff floor," and "soft mounting," areas on transmissibility-ratio surface.

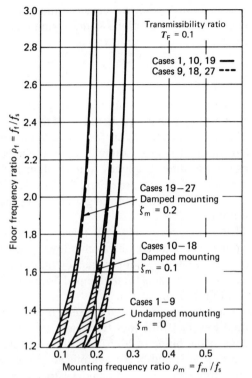

Figure 8.9. Envelopes of contours for transmissibility ratio 0.1, for systems with mass ratio 0–1.0; floor damping ratio 0–0.2, for three values of mounting damping ratio (for case numbers see Table 8.1). (Macinante and Simmons, 1976).

1.5 is to make the mounting natural frequency a small fraction of the excitation frequency, for it will then be a still smaller fraction of the resonance frequency.

Stiff Mounting

If the mounting natural frequency is much higher than the excitation frequency ($f_m/f_s > 1.5$), the behavior of the system is like that of a one-mass system. This is evident in the shape of the "end wall" where $f_m/f_s = 2$, not directly seen in Figs. 8.3 and 8.8. Because the mounting is relatively stiff the system behaves like a rigid mass (machine plus mounting) on a springy floor. The peak response is attributable to floor resonance. The position of the resonance peak varies with the mass ratio. This is indicated in Fig. 8.2 for an undamped system by the separation of the curves at the right of the illustration.

In the stiff mounting zone there is an area of low T_F in the upper corner of the illustrations in Figs. 8.3 and 8.8, which is associated with low values of the floor frequency ratio, that is, "soft floor." Design for low T_F by using a stiff mounting on a soft floor is pointless because under these conditions the mounting is redundant: the floor functions as the spring element and the floor vibration may be excessive.

Design Area

In the light of the foregoing, the "design area," in which the "design point" defined by the coordinates f_m/f_s and f_f/f_s should be placed, is the strip of "low ground" along the lower right of the T_F surface bounded by the contour of the nominated maximum value of T_F, for example, $T_F = 0.3$ as illustrated in Fig. 8.8, and by a lower limit of floor frequency ratio of about 0.5. Low values of T_F are available with $f_f/f_s < 0.5$ (e.g., see Figs. 8.6 and 8.7) but design is not recommended in the soft-floor, stiff-mounting area where the floor functions as the isolating medium.

An inspection of the surfaces in Fig. 8.3 (cases 1, 10, 19) that relate to cases in which the floor is undamped, shows that low T_F is unattainable on or near the ridge that occurs where the floor frequency ratio is unity. This is seen more clearly in the contour diagrams for undamped floors (e.g., Fig. 8.5, Case 2; Fig. 8.6, Case 11; Fig. 8.7, Case 20).

Even with a floor having appreciable damping, it may be difficult or impracticable to attain a low T_F if the floor frequency ratio is in the vicinity of unity. This is evident in the bulge of high T_F at the right side of each of the other six surfaces in Fig. 8.3, and in the contour diagrams, which show that a very low mounting frequency ratio is required to attain a low T_F for an installation having a floor frequency ratio close to unity. For example, referring to the contour diagrams for floor damping ratio $\zeta_f = 0.1$ and mass ratio $R_m = 0.5$, inspection of Figs. 8.5, 8.6, and 8.7 for Cases 5, 14, and 23, respectively, shows that in order to attain $T_F < 0.1$, the mounting frequency ratio required is in the low range of 0.05–0.14, the particular value depending on the mounting damping ratio.

"Overkill"

From the foregoing demonstration of the difficulty, perhaps impossibility, of attaining a low T_F if the floor frequency ratio is in the vicinity of unity, obviously the practice of designing on the assumption that an exaggerated static deflection can compensate for floor flexibility involves the risk that if the floor frequency ratio happens to be close to the excitation frequency, a high transmissibility will result.

With some types of isolator there is some risk also in the assumption, tacit or otherwise, that increasing the static deflection progressively lowers the natural frequency of the mounting. The effects of overloading and nonlinearity of the isolators (e.g., Example 7.2) may frustrate the intended lowering of natural frequency with increase in static deflection.

Design guides relevant to the overkill approach, which are in the form of recommended static deflections for particular types of machinery on floors of various spans, are based on many years of experience with actual installations. In using these recommendations, which promise a low mounting frequency ratio, and hence a conservative design, the designer of the mounting can avoid the risk mentioned above by estimating the floor natural frequency, and making use of design data given in this chapter to check that the floor frequency ratio is sufficiently removed from unity to give a suitably low transmissibility ratio.

PRACTICAL APPLICATION OF THE DESIGN DATA

The design data show how T_F is influenced by the design variables associated with the machine, the mounting, and the floor. In practice, the design of a plant room floor and the choice of machinery are usually decided on considerations other than vibration, leaving only the design of the mounting to be determined. The way in which the design data are used for this purpose is described in this section.

However, the design data may be used also in relation to the design of the floor and the choice of machinery. We have seen that when the floor frequency ratio is close to unity a very low value of the mounting frequency ratio may be required to achieve a sufficiently low value of T_F. This means that an exceptionally soft and relatively costly mounting may be required to ensure the desired vibration control. Therefore, if calculations in relation to a preliminary design of the floor indicate that the floor natural frequency would be in close proximity to the excitation frequency, the design of the floor and/or choice of the machinery should be modified so that these frequencies are separated.

Determine Floor Characteristics

Although the purpose of the procedure being presented is to take floor flexibility into account, the physical quantity (stiffness; force per unit displacement) that is the usual measure of flexibility is not treated as a primary design factor. Instead, the natural frequency is taken as a more meaningful parameter where the dynamic behavior of the floor is concerned. This is not inconsistent with what has been said about the importance of the floor stiffness, because the stiffness is a major factor in determining the natural frequency.

However, a numerical value of the floor stiffness may be used for the secondary purpose of assigning a value to the equivalent mass of the floor and hence the mass ratio. The calculation of the stiffness of the floor from the structural drawings is a matter for the structural engineer and is not discussed here. Biggs (1964, Chap. 5) presents methods of calculation and design data for the stiffness and mass of equivalent one-mass systems representing floors, beams and slabs supported in various ways. In relation to the design of an installation on an existing floor, the stiffness could be found experimentally, by observing the deflection of the floor under test loadings, but in practice the cost would probably not be justified. The effort that should be devoted to determination of the mass ratio may be judged by referring to the T_F contours in the part of the design area relevant to the installation, and noting the influence of variation in the mass ratio.

Natural Frequency

Computation of the natural frequency of a proposed floor from structural drawings is a matter for a specialist in structural dynamics and is outside the scope of this book. Examples of the kind of computation involved are given by Caro et al. (1975) who used methods described by Biggs (1964, Chaps. 3 and 5) to calculate the natural frequency and stiffness of 12 existing concrete floors. Caro et al. also give the actual values of the natural frequencies observed experimentally. The maximum discrepancy between any calculated and the corresponding observed value in 12 pairs in the range 2–10 Hz was about 15%.

The natural frequency of an existing floor may be found experimentally by analyzing the vibration of the floor resulting from transient excitation (e.g., dropping a bag of sand or other heavy object onto the floor). Alternatively, the floor natural frequency may be inferred from the observed spectrum of the response of the floor to ambient excitation. Another method is to use a vibration generator suitable for exciting the relatively low frequencies necessary, and adjust the frequency to produce resonance of the floor in the fundamental mode.

The floor natural frequency is the quantity most likely to be difficult or inconvenient to evaluate at the design stage. It will become obvious that the design data can be used to assess the influence on T_F of uncertainty in the value assigned to the floor natural frequency.

Mass Ratio

In the design model the floor is represented by a damped one-mass system (Fig. 8.1a). The natural frequency decreases with increased damping (e.g., see Den Hartog, 1956, p. 40), but the effect is small for low values of the damping ratio. For example, with floor damping ratio $\zeta_f = 0.2$, which is higher than that of most existing floors, ignoring the damping involves an error of about 2%. Therefore, for the purpose of calculating the equivalent mass of the floor for given values of floor natural frequency and stiffness, the damping is ignored and the relationship for the undamped one-mass system is used:

$$f_f = \frac{1}{2\pi} \left(\frac{k_f}{m_f}\right)^{1/2}, \quad \text{or} \quad m_f = \frac{k_f}{4\pi^2 f_f^2} \quad (8.11)$$

Then, R_m is the ratio m_m/m_f, where m_m is the total mass of machine, plus inertia block, if any, that is to be supported on isolators.

Damping Ratio

If vibration measurements are made to evaluate the floor natural frequency, the floor damping ratio ζ_f can be found at the same time (e.g., from transient response, see Example 2.2). There is no need to make experiments to measure only the floor damping ratio. Most installations will be on reinforced concrete floors, which can be expected to have a damping ratio less than 0.1 and commonly less than 0.05. If the installation is to be on a steel deck or platform the damping ratio will probably be 0.01–0.02. For an installation designed for operation well away from resonance ζ_f can be taken as zero, or the design data can be interpolated between $\zeta_f = 0$ and $\zeta_f = 0.1$.

Determine Excitation Frequency

Sources of periodic vibration are conveniently classified as "rotating" or "reciprocating" machines. The former group includes machines such as turbines, electric motors, and generators whose main elements are purely rotational, whereas the latter group includes internal combustion engines, reciprocating pumps and compressors, vibrators, and other machines having significant reciprocating as well as rotating elements.

The excitation that derives from static and dynamic unbalance of a rotor can be regarded as equivalent to a vertical and a horizontal oscillatory force component acting through the c.g. of the rotor, together with oscillatory couples in the vertical and horizontal planes through the c.g. of the rotor. The excitation frequency is the "once-per-revolution" component referred to in Chapter 4 under the heading Balance Quality; therefore, the design value assigned to f_s is equal to the rotational frequency of the rotor.

The excitation generated by the piston–connecting rod–crankshaft system that is the basic mechanism in most reciprocating machinery is complicated in the ways indicated in the next few paragraphs (e.g., see Crede, 1951, pp. 267–273; Den Hartog, 1956, Chap. 5; Ker Wilson, 1959, Part A).

The reciprocating mass produces an oscillatory inertial force which normally is only partially compensated by a counterweight on the crankshaft because a counterweight, while compensating in the direction of the piston stroke, introduces an unbalanced force component perpendicular to the piston stroke and to the crankshaft.

A further complication results from the fact that the motion of the piston is not purely sinusoidal. When the crank pin has turned 90° from a dead center position the piston is not at mid-stroke because of the finite length of the connecting rod; as a consequence there are second, fourth, and so on, harmonics superposed on the fundamental sinusoidal acceleration of the reciprocating mass.

Yet another contribution to the excitation results from the fact that the transfer of force between piston and connecting rod while the latter is in an angular position causes a transverse force at the cylinder which, together with the reaction to this force at the crankshaft bearings, forms an oscillatory couple tending to rock the engine about some axis parallel to the crankshaft. The rocking excitation generated by the firing stroke in each cylinder occurs at shaft frequency if the operating cycle is two-stroke, and at half shaft frequency if four-stroke.

With multicylinder reciprocating machinery fortunately the severity of the unbalanced forces and couples does not go on increasing with the number of cylinders. On the contrary, the forces and couples associated with the operation of one cylinder are compensated by those of other cylinders to an extent that depends on the number and arrangement of the cylinders and the firing sequence. For example, a six cylinder in-line engine theoretically can be perfectly balanced.

How does the designer of the mounting get the information required about the state of unbalance of the machine that is to be isolated? Presumably the machine designer would have calculated or estimated the magnitudes, frequencies, and directions of the significant forces and couples in order to design the necessary strength into the components and frame of the machine. This is the information that is required by the designer of the mounting, but which is rarely obtainable through the manufacturer or supplier of the machine. Usually the designer of the mounting must assess the excitation from a knowledge of the operating speed and the dynamics of the machine as discussed in the references cited.

The excitation frequency can be found experimentally by observing the vibration of the machine or a similar one in operation, bearing in mind that although the observed amplitudes will be influenced by the method of supporting the test machine, the observed frequency spectrum provides a valid basis for assigning a value to the excitation frequency f_s.

For the design of the mounting, usually only the lowest excitation frequency needs to be known. High frequency components at gear-meshing and blade-passage frequencies, and random excitation from bearings, belt drives, cam and valve actions usually can be ignored as far as the design of the mounting is concerned. However, with metal springs and rubber isolators the transmissibility at the particular values of the frequency in the high frequency range where standing waves occur in the isolators may be significantly higher than would be expected on the basis of theory, as presented here, which assumes the isolators to be massless linear springs (see Chapter 6, Case for Adoption of Two-Mass Model, Lumped or Distributed Properties).

Evaluate Floor Frequency Ratio

Using the values assigned to f_f and f_s, evaluate f_f/f_s. There may be several values, or a range, to be considered, taking into account the uncertainty in the estimates of f_f and f_s, and perhaps also a range in f_s associated with the operating range of the machine.

Assign Damping Ratio of Mounting

In practice, some damping is usually considered necessary to limit the response of the mounting during a resonance that may occur on starting or stopping the machine. With helical spring isolators assume $\zeta_m = 0$, unless separate dampers are provided. For air spring and rubber isolators commonly $\zeta_m \approx 0.1$, but for some rubber isolators $\zeta_m \approx 0.2$.

Identify Case Number

For the particular values of the mass ratio R_m and of the two damping ratios ζ_m and ζ_f refer to Table 8.1 and identify the case number. If alternative designs are being considered, or there are uncertainties about some of the parameters, identify any other candidate case.

Nominate Transmissibility Ratio

A low T_F means that the force amplitude transmitted is low in comparison with the amplitude at the source, but not necessarily low enough for the consequent vibration to be acceptable. The installation is successful if the level of transmitted vibration is acceptable to the occupants of the building. The criterion of acceptability depends on their activities. In a residential building the vibration should be imperceptible to those working, reading, relaxing, sleeping, or trying to sleep. In a laboratory or hospital the vibration, even though imperceptible to humans, must not disturb sensitive apparatus or operations. It need hardly be said that, in all buildings, the level of vibration must be below that which can cause damage to the structure. Unfortunately, there is no simple relationship between the T_F as defined and acceptable vibration as discussed in Chapter 4.

The aim in the design of the mounting is to provide a "low enough" value of T_F. For practical purposes a nominal value of $T_F = 0.1$ is suggested, but if machinery is to be installed close to a critically vibration-sensitive area, a lower T_F should be considered. Whether or not the residual or transmitted vibration will be acceptable is a matter for separate consideration.

Evaluate Mounting Natural Frequency

For a particular case number refer to the appropriate set of T_F contours. For the particular value of floor frequency ratio f_f/f_s read the value of the mounting frequency ratio f_m/f_s necessary to achieve the desired value of T_F. From this and the known value of f_s, calculate the required value of f_m. The value of f_m

that is required will suggest the type of isolator likely to be suitable: rubber isolators for f_m down to, say, 10 Hz, helical springs to about 3 Hz, and air springs to 1 Hz and lower, as indicated in the following example.

Example 8.2. A machine operating at 960 rev/min is to be installed on a suspended floor, on a seismic mounting designed so that the alternating force transmitted into the structure supporting the floor does not exceed 10% of that generated by the machine. Determine the vertical natural frequency that should be specified for the mounting. Details of the installation are as follows:

Major alternating force generated by machine occurs at shaft frequency; that is, $f_s = 16$ Hz

Mass of machine plus inertia block $m_m = 4800$ kg

Floor vertical stiffness $k_f = 38$ kN/mm (38×10^6 N/m)

Floor natural frequency $f_f = 10$ Hz

Floor damping ratio $\zeta_f = 0.1$.

Floor mass: for the given values of the floor stiffness and natural frequency, the equivalent mass of the floor is found from Eq. (8.11)

$$m_f = \frac{k_f}{4\pi^2 f_f^2} = \frac{38 \times 10^6}{4\pi^2 \, 10^2} \approx 9620 \text{ kg}$$

Mass ratio $R_m = m_m/m_f = 4800/9620 \approx 0.5$

Floor frequency ratio $f_f/f_s = 10/16 = 0.62$

For the particular values $R_m = 0.5$ and $\zeta_f = 0.1$ the installation is Case 5, 14, or 23 depending on the value of the mounting damping ratio ζ_m, which is not known until the type of isolator is decided.

First consider using rubber isolators having $\zeta_m = 0.2$. This is a Case 23 installation. Refer to Fig. 8.7 and use the broken line contours that relate to Case 23. Using the $T_F = 0.1$ contours read that for $f_f/f_s = 0.62$ the required value of f_m/f_s is 0.28 (point A). Since $f_s = 16$, the required value of f_m is 0.28 $\times$ 16 = 4.5 Hz. This is below the range of f_m normally attainable with rubber isolators.

Now consider the use of helical spring isolators, for which the damping is negligible, $\zeta_m = 0$. This is a Case 5 installation. Refer to Fig. 8.5 and use the contours for Case 5. For $f_f/f_s = 0.62$ and $T_F = 0.1$, read that the required value of $f_m/f_s = 0.35$ (point B). Therefore, $f_m = 0.35 \times 16 = 5.6$ Hz. This is easily attainable with helical spring isolators. In fact, f_m as low as 3 Hz is quite common. With $f_m = 3.0$, that is, with $f_m/f_s = 3/16 = 0.19$ (point C), T_F is seen to be less than 0.03.

Alternatively consider the use of air springs having $\zeta_m = 0.1$. This is a Case 14 installation. Refer to Fig. 8.6 and use the contours for Case 14. For

$T_F = 0.1$ and $f_f/f_s = 0.62$ the required $f_m/f_s = 0.32$ (point D). That is, $f_m = 0.32 \times 16 = 5.1$ Hz, which is easily attained with air spring isolators.

For a critical installation, air springs providing $f_m = 1$ Hz might be considered. With $f_m/f_s = 1/16 = 0.06$ (point E), the very low value $T_F = 0.01$ is attainable.

Select Isolators and Arrange Layout

The foregoing procedure evaluates the mounting natural frequency that will give the required quality of isolation. To achieve this desired natural frequency in the hardware, nominate a convenient number of isolators and arrange them in such a way that the mounting has an uncoupled vertical mode (see Example 7.1). Then in the manner outlined below choose isolators of stiffness such that the natural frequency in this mode has the required value.

The natural frequency in the uncoupled vertical mode is determined by the seismic mass and the total vertical stiffness of the mounting, as shown by Eq. (7.2). In using this equation in the present context, note that Eq. (7.2) gives the natural frequency of a one-mass system, and the notation is appropriate to that context. Thus, in Eq. (7.2), the natural frequency (f_m in the context of the two-mass system) is denoted by f_z, the seismic mass (m_m) is denoted by M, and the total vertical stiffness of the mounting (k_m) is denoted by K_z.

The required value of the total vertical stiffness of the mounting is found by substituting into Eq. (7.2) the desired natural frequency of vertical vibration and the estimated seismic mass. The required stiffness (k_z) of an individual isolator is found by dividing the total stiffness by the number of isolators.

It is now necessary to select, from catalogs of commercially available isolators of a suitable type, an isolator of a suitable load rating, having stiffness of the required value, remembering that the stiffness k_z is the dynamic stiffness under the working conditions of load and frequency, as described in Chapter 5 under the heading Rubber Isolators and illustrated in Example 7.2.

Some trial and compromise may be necessary in deciding the particular isolator, because isolators in a commercial range of a given type are manufactured only in a limited number of stiffness ratings. An isolator is chosen having stiffness nearest to the required value, and the natural frequency is calculated for the nominated number of isolators. The design data are then used "in reverse" to find the T_F promised by this arrangement. If it is unacceptable, other combinations of number and stiffness of isolator are tried until a satisfactory combination is found.

Check Vibration Amplitudes

Having arrived at a design that promises low force transmissibility, the designer should check that the vibration amplitudes of machine and floor are within acceptable limits (Chapter 4).

The amplitudes can be calculated by using the general expressions which are referred to under the heading Basis of Computations, but are not detailed for the reasons given. The calculation involves the substitution of the values of five design variables (two frequency ratios, two damping ratios, and the mass ratio) into six groups or packages of algebraic terms, then using the results to evaluate four more groups, and finally using the latter results to evaluate the amplitude ratios, from which the amplitudes are found. Although this calculation can be done with patience and a pocket calculator, obviously a programmable computer is preferable.

As mentioned earlier under Vibration Amplitudes the floor amplitude ratio can be derived without recourse to the general expression if the transmissibility ratio is specified or nominated, as it is in the normal design situation (see Example 8.3). Also in the earlier context it is mentioned that the damping has its major influence on the response amplitudes under resonance conditions: a numerical illustration of this is included in Example 8.3.

The designer must not lose sight of the fact that the amplitudes calculated by the methods described are specifically those of the vertical vibration of the two masses in the design model. In an actual installation the amplitudes of vibration of machine and floor may differ significantly from the calculated values for the reasons indicated in the following paragraphs.

Machine Vibration Amplitude

The vertical vibration of the seismic mass may be coupled with rotation about a horizontal axis; consequently, the vertical component of vibration of parts of the machine and inertia block remote from the c.g. of the seismic mass may be significantly greater than the calculated amplitude.

If the machine also generates forces and couples which cause horizontal vibration of the seismic mass coupled with angular vibration about a vertical axis, the resultant of the horizontal and vertical components of vibration of parts of the machine and inertia block remote from the c.g. may be well in excess of the calculated amplitude.

Floor Vibration Amplitude

The floor vibration amplitude is calculated on the assumption that the floor behaves like a lumped mass–spring–damper system having the same natural frequency, stiffness, and damping as the floor. In a practical situation the installation may not be located over the center of the floor, the floor stiffness may be nonuniform, and the machine may generate forces and couples that, when transmitted through the isolators, tend to excite the second and higher modes of floor vibration rather than, or in addition to, the fundamental mode.

Because of the relatively high stiffness of the floor in its plane, the effects of horizontal excitation may be ignored insofar as the floor itself is concerned.

However, if the plant room is at an upper level in a tall building, the excitation applied at the floor may cause significant vibration of the building as a whole in its flexural and torsional modes as a vertical cantilever.

Example 8.3.   An installation comprising a 500 rev/min machine, seismically mounted on the floor referred to in Example 8.2, is operating in the resonance region represented by the crest of the T_F surface for Case 13 (Fig. 8.3e). The "design point" is defined by floor and mounting frequency ratios each 1.2. In the illustration it is evident that the design point is above the $T_F = 10$ contour; the calculated value is given as $T_F = 11.4$. Calculate the floor vibration amplitude and compare the result with the value calculated ignoring the system damping. The following data refer to the installation.

The primary excitation is at shaft rotational frequency ($f_s = 8.3$ Hz), resulting from unbalance of a 310 kg rotor balanced to ISO quality grade G6.3 (125 g·mm/kg).

Seismic mass $m_m = 960$ kg
Mass ratio $R_m = 0.1$
Mounting frequency ratio $\rho_m = f_m/f_s = 1.2$
Mounting natural frequency $f_m = 1.2 \times 8.3 = 10$ Hz
Mounting damping ratio $\zeta_m = 0.1$
Mounting stiffness $k_m = 3.8$ kN/mm
Floor frequency ratio $\rho_f = f_f/f_s = 1.2$
Floor natural frequency $f_f = 1.2 \times 8.3 = 10$ Hz
Floor damping ratio $\zeta_f = 0.1$

For the given value $T_F = 11.4$ the floor amplitude ratio Z_f/δ_P is found by using Eq. (8.3) as follows:

$$T_F = \frac{Z_f}{\delta_P} \frac{\rho_f}{R_m \rho_m^2} (\rho_f^2 + 4\zeta_f^2)^{1/2}$$

$$11.4 = \frac{Z_f}{\delta_P} \frac{1.2}{0.1(1.2)^2} [(1.2)^2 + 4(0.1)^2]^{1/2}$$

which reduces to

$$\frac{Z_f}{\delta_P} = 1.12$$

To evaluate the amplitude Z_f we need the value of δ_P, the deflection of the mounting under a force equal to the amplitude P_S of the applied alternating force. In this example P_S is the vertical component of the centrifugal force resulting from rotor unbalance. The given value, 125 g·mm/kg of rotor mass (310 kg), represents a mass radius product of $125 \times 310 \times 10^{-6}$ kg·m. The rotational frequency is 8.3 rev/s, that is, $\omega_s = 2\pi \times 8.3$ rad/s. Therefore,

$$P_S = mr\omega_s^2 = 125 \times 310 \times 4\pi^2(8.3)^2 \times 10^{-6} \text{ N} = 106 \text{ N}$$

and

$$k_m = 3.8 \text{ kN/mm}$$

Therefore,

$$\delta_P = \frac{P_S}{k_m} = \frac{106}{3800} = 0.028 \text{ mm}$$

and

$$Z_f = 1.12 \, \delta_P = 1.12 \times 0.028 = 0.031 \text{ mm (31 } \mu\text{m)}$$

If the damping is ignored, the floor amplitude is calculated from Eqs. (8.4) and (8.7) as follows:

$$\frac{Z_f}{\delta_P} = \frac{R_m \rho_m^4}{D}$$

where $D = 1 - (R_m + 1)\rho_m^2 - \rho_f^2 + \rho_m^2\rho_f^2$

$$= 1 - (0.1 + 1)(1.2)^2 - (1.2)^2 + (1.2)^2(1.2)^2$$

$$= 0.05$$

Therefore,

$$\frac{Z_f}{\delta_P} = \frac{0.1(1.2)^4}{0.05} = 4.15$$

and

$$Z_f = 4.15 \, \delta_P = 4.15 \times 0.028$$
$$= 0.116 \text{ mm (116 } \mu\text{m)}$$

This floor amplitude calculated without taking the damping into account is much greater than the value (31 μm) calculated for the damped system.

As a matter of interest we may note the influence of the damping on the transmissibility ratio and the machine vibration amplitude. The author's calculations from the general expressions taking damping into account give:

$$\text{Transmissibility ratio, } T_F = 11.4$$

$$\text{Mounting amplitude ratio, } \frac{Z_m}{\delta_P} = 4.9$$

Therefore,

$$Z_m = 4.9\delta_P = 4.9 \times 0.028$$
$$= 0.137 \text{ mm } (137 \text{ } \mu\text{m})$$

Ignoring the damping, the transmissibility ratio calculated from Eqs. (8.6) and (8.7) is

$$T_F = \frac{\rho_m^2 \rho_f^2}{D}$$

$$= \frac{(1.2)^2(1.2)^2}{0.05}$$

$$= 41.5$$

This is much higher than that of the damped system, which is understandable because by ignoring the damping the Case 13 system is being treated as Case 1, for which the transmissibility ratio in the resonance region is obviously much higher (see Fig. 8.3a). It may be noted in Fig. 8.2 that the design point $\rho_m = \rho_f = 1.2$ for Case 1 is not far removed from the curve representing the conditions under which T_F tends to infinity.

Ignoring the damping in calculating the machine vibration amplitude, Eqs. (8.5) and (8.7) give

$$\frac{Z_m}{\delta_P} = \frac{R_m \rho_m^4 + \rho_m^2 \rho_f^2 - \rho_m^2}{D}$$

$$= \frac{0.1(1.2)^4 + (1.2)^2(1.2)^2 - (1.2)^2}{0.05}$$

$$= 16.8$$

Therefore,

$$Z_m = 16.8 \text{ } \delta_P = 16.8 \times 0.028$$
$$= 0.470 \text{ mm } (470 \text{ } \mu\text{m})$$

This is much higher than the calculated value (137 μm) taking damping into account.

CONCLUDING REMARKS

The design data presented in this chapter are derived from computations based on a theoretical analysis of the behavior of a damped two-mass system which is taken as the design model of the installation that is being designed. The validity of the design data in practical applications depends on the adequacy of the design model and assumed excitation as a representation of the particular application, bearing in mind the limitations imposed by the simplifying assumptions about the excitation and the dynamic characteristics and behavior of the model as discussed in Chapter 6.

The main limitations are that consideration is given only to the vertical component of the alternating force generated by the machine, that the machine plus inertia block is treated as a single lumped mass, that the floor vibrates in its fundamental flexural mode and is represented as a simple mass–spring system, that the center of gravity of each of the two masses lies on the vertical line of action of the alternating force, and that the vibration of each mass is purely vertical. Clearly, this leaves the designer with the responsibility of considering and if necessary designing to minimize the response of the system in modes other than vertical.

One should not put too much trust in any design model until its adequacy is confirmed in the performance of the hardware. Those involved in the design of mountings for machinery on suspended floors should take any opportunity to determine the behavior of the installation by vibration measurements, and to compare the observed vibration with that calculated for the two-mass model. From such comparisons it will be possible to build up a background of experience of the types of installation for which the two-mass model is adequate.

REFERENCES

American Society of Heating, Refrigerating and Air-Conditioning Engineers (1980). Sound and Vibration Control, in *ASHRAE Handbook & Product Directory, 1980 Systems*, Chap 35.

Biggs, John M. (1964). *Introduction to Structural Dynamics*, McGraw-Hill, New York.

Caro, J. C., G. K. Larkins, and T. H. Cairnes (1975). *Vibration in Prestressed Slabs with Particular Reference to Hospital Structures*, Symposium on Serviceability of Concrete, Institution of Engineers, Australia, National Conference Publication, No. 75/6, pp. 34–40.

Crede, Charles E. (1951). *Vibration and Shock Isolation*, Wiley, New York.

Crist, R. A., and J. R. Shaver (1976). *Deflection Performance Criteria for Floors*, U.S. Bureau of Standards, NBS Tech. Note 900.

Den Hartog, J. P. (1956). *Mechanical Vibrations*, 4th ed., McGraw-Hill, New York.

Eason, A. B. (1919). Critical Speeds of Machinery Placed on Upper Floors of Buildings, as Related to Vibration, *Phil. Mag.*, **38**, 6th Series, 1919(2), 395–402. (See also Eason 1923, Appx. II, 149–155.)

Eason, Alec B. (1923). *The Prevention of Vibration and Noise*, Henry Frowde and Hodder & Stoughton, London.

Eberhart, L. L. (1966). *ASHRAE J.*, **8**(5), 54–60.

Galambos, T. V., P. L. Gould, M. K. Ravindra, H. Suryoutomo, and R. A. Crist (1973). *Structural Deflections. A Literature and State-of-the-Art Survey*, U.S. Bureau of Standards, NBS Building Science Series 47.

Ker Wilson, W. (1959). *Vibration Engineering*, Charles Griffin, London.

Macinante, J. A., and H. Simmons (1976). Design Criteria for Vibration Isolating Mountings for Machinery on Suspended Floors, *Vibration and Noise Control Engineering Conference*, Institution of Engineers, Australia, National Conference Publication No. 76/9, pp. 46–50.

Macinante, J. A., and H. Simmons (1977). Vibration Isolating Mountings for Machinery on Suspended Floors, *Inst. Eng. Aust. Mech. Eng. Trans.*, **ME2**, 27–35.

Macinante, J. A., and H. Simmons (1981). *Vibration Isolating Mountings for Machinery on Suspended Floors Design Data No. 1*, Institution of Engineers, Australia.

Snowdon, J. C. (1968). *Vibration and Shock in Damped Mechanical Systems*, Wiley, New York.

Steffens, R. J. (1966). Some Aspects of Structural Vibration, *Vibration in Civil Engineering Symposium*, London, Butterworths, London, pp. 1–30.

Stodola, A. (1905). *Steam Turbines*, Van Nostrand, New York, pp. 355–358.

U.S. General Services Administration (1971). Public Buildings Service *Guide Specification, Vibration Isolation*, PBS 4-1515-71.

9

Mountings for Sensitive Equipment

The danger of more or less perpetual vibration of significant magnitude is one of the bugbears of designers of accurate instruments, and research leading to some practical data on this subject for various types of members is urgently required.

T. N. WHITEHEAD (1934, p. 184)

Today, 50 years after Whitehead wrote these words, vibratory disturbance of important parts of sensitive equipment is still a bugbear of the designer and user of the equipment. The responsive part may be, for example, the optical head of a measuring or photographic equipment, the stage holding the specimen in an electron microscope, or the wheelhead of a grinding machine. Whatever it is, for brevity we call it the *critical element*. The various kinds of disturbance (e.g., blurred optical image, ripple marks on ground surface), which are symptoms of unwanted vibration of critical elements, are mentioned in Chapter 4 under the heading Sensitive Equipment.

The critical element may be an overhung or cantilevered member of the kind sketched in Figs. 4.5 and 6.7, or a bracket supporting a lens or mirror in an optical system. It may be a suspended or pendulous element, such as the torsional suspension of a galvanometer, or a pool of mercury whose surface forms part of an optical system. Often the critical element is unidentified and inaccessible within the equipment. For the purposes of this chapter we assume that the critical element is a structural member having the properties of mass, stiffness, and damping.

There may be more than one such element and the important consideration the vibratory displacement of the elements relative to one another or to the base of the equipment. A critical element may vibrate in more than one mode and, to complicate matters further, it may be in contact with another part of the equipment as, for example, the wheelhead of a grinding machine is in contact at the wheel–workpiece interface. At present we leave aside all these complications, and assume that there is only one critical element, which behaves as a simple mass–spring–damper system attached to the body of the equipment.

Vibration of the critical element may result from excitation applied inadvertently by the operator during manipulation or adjustment of the equipment, or from the operation of accessories such as pumps and motors on the equipment itself, but usually the major source of disturbance is site vibration received through the structure supporting the equipment.

In response to a transient excitation the critical element makes a free vibration at its own natural frequency. The higher the natural frequency the smaller the displacement amplitude, and the greater the damping the more rapid the decay of the transient.

The critical element responds to steady-state excitation with a vibration whose frequency is that of the excitation. The amplitude is determined by the damping and the ratio of the forcing to the natural frequency. The response of the critical element relative to the base of the equipment is of the form shown in Fig. 4.5b, which differs from the response curve in Fig. 6.5 because the latter shows the *absolute* displacement transmissibility ratio, whereas Fig. 4.5b shows the *relative* displacement transmissibility ratio. Thus when the excitation frequency is a very small fraction of the natural frequency, the displacement transmissibility ratio is unity in Fig. 6.5, which means that the displacement amplitude of the critical element is equal to that of the base of the equipment. Because these displacements are in phase, the relative displacement amplitude of the critical element is zero, as shown in Fig. 4.5b.

The response curve in Fig. 4.5b is that of a critical element with natural frequency f_c = 20 Hz (Example 4.2). With low values of the site vibration frequency, for example, f_s = 10 Hz, f_c is significantly higher than f_s and the displacement of the critical element relative to the frame of the equipment is small. For high values of site vibration frequency, for example, f_s = 60 Hz, f_c is much lower than f_s and the system is basically that of a vibrometer or displacement measuring device (Chapter 2, Vibration Measurement), in which the relative displacement of the element is about the same as the absolute displacement of the support. If f_s is about equal to f_c the system amplifies the vibration, and the relative displacement of the critical element is greater than the support displacement.

From what has been said about the response to transient and steady-state excitation, obviously the critical element should have a high natural frequency—but how high? Whitehead, in the context of the opening quotation, saw the need for individual components to have high natural frequencies, and suggested 200 Hz as a value to be aimed at. Crede (1951, p. 148) mentioned that the natural frequencies of critical elements in a vacuum tube or surface grinder are often 100 Hz and higher.

CRITICAL-ELEMENT FACTOR IN DESIGN OF MOUNTING

If, in fact, the critical element has a natural frequency of 100 Hz or higher, which is well above the usual ranges of both site vibration frequency and mounting natural frequency, then one might well assume that the critical element can be ignored as a factor in the design of the mounting.

However, the critical elements of large items of sensitive equipment marketed by reputable manufacturers can have much lower natural frequencies. In the course of experimental investigations during the period 1955–1965 on six large machine tools (three roll grinders, two gear grinders, one jig borer), the author observed 15 values of the natural frequencies of critical elements [e.g., grinding-wheel head, workpiece, tool spindle (see Examples 5.12)] in the range 18–46 Hz. Rivin (1979a, Table 1) lists 26 values of "resonance frequency of upper system" (i.e., secondary system in two-mass model) of a variety of precision machine tools, in the range 28–100 Hz.

Perhaps the reason why a critical element sometimes has a low natural frequency is that the designer does not recognize the natural frequency as an important design factor. Traditionally and intuitively the designer makes the member "look right" on the drawing board, and perhaps checks its deflection under a static force estimated to be equivalent to its loading in service. This can give a misleading impression of the adequacy of the design from the viewpoint of the dynamic response, for which the natural frequency is more meaningful. A component may look very stiff because of its robust proportions and generous cross section, but it is also massive and its natural frequency may be surprisingly low, because of the influence of the mass and mass distribution, and the compliance of bolted joints, guides, and bearings connecting the critical element to the body of the equipment.

When equipment having a critical element of low natural frequency is disturbed by vibration the user, perhaps not realizing that the equipment has an exceptionally responsive critical element, or taking the view that the equipment cannot be modified to reduce its response, usually decides to provide a seismic mounting. The characteristics of the critical element are rarely considered in relation to the mounting, which usually is designed on the basis of the one-mass model (Chapter 6) to satisfy the simple criterion that the natural frequency should be low in comparison with the predominant frequency of the site vibration.

The extent to which the use of the one-mass design model has resulted in unsatisfactory or mediocre mountings is not known because seismic mountings are rarely tested. This is understandable because it is difficult, and usually impracticable, to make a meaningful test (e.g., see Chapter 5, Performance Testing). A demonstration of satisfactory performance of the equipment is not proof that the mounting is effective; the equipment may have performed as well or better without the mounting.

No doubt many mountings designed on the basis of the one-mass model prove successful because of fortuitously high natural frequencies of the critical elements, whereas others are unsatisfactory because the natural frequencies of critical elements are low enough to be in the vicinity of site excitation frequencies. If also, because of inadequate design or performance of the mounting, its natural frequency is close to the excitation frequency, the site vibration is magnified first by the mounting and then by the critical element. Although this may seem like the kind of special case that the theorist feels obliged to examine "for the sake of completeness," it can happen in practice. Indeed, it was such a coincidence of frequencies (see Example 5.5) that stimulated the author to investigate the influence of the critical element on the transmissibility and hence on the design of mountings.

The results of the investigation are given in this chapter. Design data are presented showing how the relative vibratory displacement of the critical element is determined by the design variables associated with the critical element, the mounting and the excitation.

The designer of the mounting can use these data to find what value should be specified for the natural frequency of a mounting for an equipment having a critical element of given natural frequency and damping, to attain a nominated small response of the critical element when the base of the equipment is subjected to a given damped harmonic or steady-state excitation.

The equipment designer can use the data to determine the minimum natural frequency of the critical element that will limit its relative displacement to a certain acceptable level, under given conditions of mounting and excitation.

THEORETICAL BASIS

The response of a two-mass system to base excitation has been studied in relation to various practical applications including the design of cushioning for the transport of fragile articles, the isolation of electronic and other sensitive equipment in land, sea, air, and space vehicles, the protection of equipment in structures subjected to earthquakes, as well as that now under consideration, the design of mountings for sensitive equipment in buildings.

The general theory that is adapted to the needs of particular applications is well documented in text and reference books on shock and vibration; see, for example, Morrow (1963) and Snowdon (1968). By way of background to the theory of the method to be presented, we briefly review some investigations that are relevant in that they involve consideration of the response of a particular part or element relative to the base or body of a seismically mounted equipment when the support is subjected to transient or steady-state vibration.

The earliest reference to the use of a subsystem to represent a flexible element appears to be that of Mindlin (1945) who presented a comprehensive and detailed study of the design of cushioning (isolators) to protect a packaged article from damage that could occur if the package is dropped. Mindlin's analysis shows how the responses of the body of the article and its critical element are influenced by the characteristics of the cushioning and those of the critical element. Since Mindlin's paper, the theory and practice of the shock and vibration isolation of packaged articles have been examined and developed by Crede (1951, Chap. 3), Mustin (1968), Hatae (1976), and others.

Crede (1951, pp. 145–149) used a two-mass model to investigate the response to site vibration of a typical element of an equipment on a seismic mounting. Crede's approach is indirect in that he begins by assuming that there are no isolators. The primary system represents the floor with the base of the equipment rigidly attached to it; the secondary system represents the critical element. Crede derives an expression for the response of this system to an impulse applied to the floor. The physical quantity of interest is the displacement of the critical element relative to the base of the equipment. Crede shows that large values of element deflection can occur when the natural frequency of the critical element coincides with that of the floor system, and concludes that the maximum deflection of the critical element can be reduced by increasing the mass of the equipment, or reducing the natural frequency of the floor.

On this basis Crede (1951, p. 148) states that an effect equivalent to decreasing the natural frequency of the floor is attained by inserting relatively flexible isolators between the equipment and the floor, but that a complete analysis of such a system becomes more laborious than the results justify. Crede continues that for practical purposes a very significant reduction in maximum deflection is attained if the mounting natural frequency is 0.1–0.2 times the natural frequency of the critical element of the equipment. Thus if the natural frequency of the critical element is 100 Hz or more, a mounting natural frequency of 10–20 Hz is very effective. For machine tools in general, Crede (1951, p. 299) recommends a natural frequency of 20–25 Hz in the vertical translatory mode. However, the assumption on which Crede's (1951) recommendations are based, that the natural frequency of the critical element is 100 Hz or more, is not tenable in view of the observations, mentioned in the preceding section, of natural frequencies of some critical elements in the vicinity of 20 Hz.

Kaminskaya (1964a) describes an experimental investigation of the response of the critical element of a machine tool that is subjected to floor vibration, and of the way this response is influenced by vibration isolators placed under the machine. The critical element concerned is the grinding-wheel head of a horizontal surface grinding machine. Tests were made with floor vibration produced by a centrifugal vibrator and by dropped weights, and the tests were repeated with the machine supported on different kinds of isolator. The response was observed in terms of physical quantities relating to two measures or indicators of the effectiveness of the isolators: the quality (surface finish, accuracy, waviness) of the machined surface; and the displacement of the cutting element relative to the workpiece. This latter quantity was measured under the same conditions of floor vibration that applied when machining the test specimens, but with the machine idle, because with the machine in operation other sources of vibration may mask the influence of floor vibration. The results indicate that a definite qualitative correlation exists between the observed surface quality and the relative displacement, and that therefore the relative displacement, with the machine idle, can be regarded as a criterion of the effectiveness of the isolators.

On the assumption that the coupling is slight between the machine bed and the critical element, Kaminskaya (1964a, p. 11) states that the effectiveness of the isolators can be estimated by considering separately the transmissibility of vibration from floor to machine bed and from machine bed to cutting zone. On this basis Kaminskaya (1964b), and Kaminskaya and Rivin (1964) present a theoretical analysis of the transmissibility of a two-mass model of a machine tool on a seismic mounting. The cutting head, which is supported on a column fixed to the bed of the machine, is represented as a secondary system with rocking response relative to the primary system, which represents the base of the machine supported on isolators. The transmissibility is defined as the ratio of the amplitude of the relative displacement of the critical element to the amplitude of the vertical vibration of the floor. This transmissibility is expressed as the product of the two separate transmissibilities: floor to machine bed, machine bed to cutting zone. On this same theoretical basis Rivin (1979b, Fig. 2) illustrates the role of the seismic mounting in reducing the response of the critical element to an acceptable level. The foregoing relates to steady-state excitation. Expressions for the response to damped harmonic and half-sine pulse excitation are included in a paper by Kaminskaya (1966) which deals with the response of machine tools to sources of internal and external mechanical shock.

Another area in which the response of critical elements has received attention is in the design of electronic equipment to withstand shock and vibration. In the opening paragraphs of his book, Crede (1951) uses the example of a vacuum tube to illustrate the need to use isolators to prevent malfunction resulting from unwanted relative vibration of the critical elements—in this case the filament, grid, and plate in the tube.

With the development and general use of solid-state devices and the demise of the vacuum tube, attention was concentrated on design to minimize unwanted response of individual components such as diodes, transistors, resistors, and capacitors, which are responsive as masses flexibly attached to the base or chassis of the equipment. As a result of studies of the response of lumped-mass models incorporating subsystems representing these flexibly attached elements, much progress was made in the theory and practice of the design of electronic equipment to minimize malfunctioning and damage that might result from shock and vibration, particularly in aircraft and space vehicles. For example, a review of technology developed at NASA, and its application to electrical systems, is given by Eshleman (1972). Basic methods for the analysis of the vibratory response of electronic equipment are presented by Steinberg (1973), who models typical electronic components such as resistors, capacitors, and diodes, supported in various ways, as beams or other structural forms, to determine their natural frequencies as critical elements. The theory and practice of equipment design to withstand shock and vibration are discussed and illustrated by Fischer (1976), and Fischer and Forkois (1976). More recently, with the application of integrated circuitry, relatively large and complicated assemblies of components supported by soldered wires, brackets, fasteners, or other means are being replaced by "chips" of trivial size, and consequently innumerable responsive critical elements are disappearing.

With the objective of evolving a reliable basis for the vibration testing of business machines, Skinner and Zable (1978) investigated the nature and magnitude of shock and vibration experienced by business machines in shipping and relocation, and in service. Vibration in service includes that caused by adjacent machinery and that generated by the machine itself. The vibratory response of the machine is determined by analysis of a design model with two subsystems representing elements flexibly attached to the body of the machine; one passive, the other carrying a source of vibration (e.g., a blower). The body of the machine on its casters (isolators) is represented as another mass—

spring system, while the floor also is treated as a mass–spring system. The response to sinusoidal excitation applied externally at the floor, or internally at one of the subsystems, is determined by solving seven simultaneous equations involving the vertical displacement of various points in the system and the angular displacement of the main mass. Conclusions are drawn about the influence of machine and element mass on the acceleration levels of machine and floor.

The two-mass and more refined models have been used in studies relating to the "seismic qualification" of equipment intended for installation in nuclear and geothermal power plants. The important consideration is the assurance of the structural integrity of essential equipment such as pumps, valves, control devices, and piping under earthquake-induced loading. For example, O'Rourke and Kountouris (1979), assuming the mass of the equipment to be small in comparison with that of the floor, determine the response of the floor to a given ground acceleration time history. The floor response is then taken as the excitation of a subsystem representing the equipment, to determine the response of the equipment.

This and other design processes are difficult. Kelly and Chitty (1980) discuss the alternative approach of protecting the equipment by constructing the entire power plant on a base isolating system. They examine the feasibility of this by testing a 20-tonne structural model of the plant on elastomeric bearings. The hardware model contains simple cantilever elements simulating equipment installations having natural frequencies corresponding to the first three natural mode frequencies of the model structure in its fixed-base configuration. The experimental results show that equipment with a natural frequency close to one of the natural frequencies of the primary structure in which it is housed can experience accelerations several times greater than those of the primary structure, and that base isolation can reduce the acceleration experienced by a sensitive equipment as a result of an earthquake.

Returning now to seismic mountings for sensitive equipment, the author's study of the practical application of the two-mass design model for seismic mountings began with observations of the response of critical elements during investigations of various machine tool installations, some of which are referred to in Chapter 5 (Examples 5.1, 5.2, 5.5, and 5.12). Concurrently with this experimental work, the author and colleague James Waldersee used a small analog computer to investigate the relative-displacement response of the two-mass system to damped harmonic excitation, and Waldersee (1960) derived mathematical expressions for the response of the system. Subsequently, when a digital computer became available, a new approach was made, taking advantage of the computer capability of determining rapidly the response of the two-mass system with various excitations and wider ranges of the design variables. The following description of these investigations is based on results published by Macinante and Walter (1973), and Macinante and Simmons (1975). A digest of these two papers is given by Macinante (1976).

Design Model

The design model, which is illustrated in Fig. 6.7 in the context of a general discussion of the two-mass model, is reproduced in Fig. 9.1 for convenience of reference and with the annotation now required. The system is assumed to have lumped properties, linear stiffness, and viscous damping as discussed in Chapter 6. The critical element is represented by a subsystem assumed to have a single degree of freedom of vertical motion. In the following notation the subscript "c" denotes "critical element," and "m" denotes "mounting."

m_c equivalent mass of critical element; that is, mass in a single degree-of-freedom system having the same natural frequency and stiffness as the critical element

m_m mass of equipment, plus workpiece or test object, plus inertia block if any, not including mass of critical element

k_c vertical stiffness of critical element

k_m vertical stiffness of mounting

c_c equivalent viscous damping coefficient of critical element

c_m equivalent viscous damping coefficient of mounting

In applying this model to a machine tool installation, the difficulty arises that when the machine is in operation the secondary mass is in contact with the primary mass (e.g., grinding wheel in contact with workpiece), but in the model the secondary mass is not in contact with the primary mass. It cannot be asserted that the relative displacement between cutting element and workpiece caused by a certain site vibration is not influenced by the contact that occurs during normal operation. An analysis of the influence of external vibration on the surface finish produced under normal cutting conditions would be exceedingly

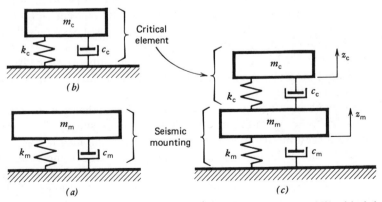

Figure 9.1. Subsystems representing (a) seismically mounted equipment and (b) critical element. (c) Two-mass model of sensitive-equipment installation.

complicated. The surface finish in grinding, for example, is influenced by such factors as the contact pressure and relative rotation of wheel and workpiece, the relative vibration of wheel and workpiece generated by the machine itself (motors, drives, pumps, etc.), vibratory phenomena (e.g., chatter) associated with the cutting action, and the influence of the width of the wheel which partly overlaps the previous cut with the result that the surface previously cut has some influence on the subsequent cut.

The application of data derived by assuming no contact, in a situation where there is contact, involves the assumption that the magnitude of the relative displacement of the critical element with no contact and machine idle, resulting from a given site vibration, may be regarded as a measure of the magnitude of the imperfection that would result from the same site vibration with the machine in normal operation.

In support of this assumption we may note that in finish grinding, for example on a roll grinder, the final traverses ("sparking out") are made with the effective depth of cut diminishing to zero. At a certain stage in the finishing operation, no further depth of cut is set, and metal removal in the final traverses results from relief of the contact pressure. Also, as mentioned earlier in this section Kaminskaya (1964a), on the basis of an experimental investigation, concluded that the relative displacement of tool and workpiece on an idle machine could be regarded as a sufficiently accurate criterion of the effectiveness of vibration isolation. In the application of the two-mass model to mountings for a wide range of sensitive equipment other than machine tools, the no-contact assumption obviously needs no defense.

Excitation

Vibration measurements at the site of a sensitive equipment installation typically indicate steady-state vibration from engines, compressors, and other continuously running machinery, transients resulting from the operation of forging hammers, punch presses and other shock sources, and random vibration.

The data to be presented relate to the response of the design model in Fig. 9.1 to two basic forms of excitation: damped harmonic and steady-state. The excitation is considered to act vertically at the surface supporting the isolators. Vibration from sources within the equipment such as motors, blowers, or other accessories is not considered.

The damped harmonic (Fig. 2.8) is an idealization of the transient vibration that results from the operation of a mechanical shock source. Whatever the shape of the force/time pulse at its origin (e.g., forging hammer anvil), the response of ground and structures in the vicinity, because of their elastic and damping properties, is a vibration having a recognizable predominant frequency, persisting for a number of cycles with decaying amplitude (Chapter 3, Transmission Through the Ground, Transmission Through Structures).

For most installations steady-state vibration is significant only at frequencies above a few hertz. However, for certain installations difficulties result from site movements of much lower frequencies. For example, Gray et al. (1972, p. 45) discuss earth motions and their influence on the behavior of inertial instruments during performance testing. They refer to two categories of earth motion. The first includes industrial and seismic vibrations that can be described by elastic wave propagation theory. The second category includes long-period nonpropagating motions, involving local and regional tilt and other phenomena, which are described by the theory of statics. They mention two basic kinds of isolating system for minimizing the motion-induced errors in the testing and operation of inertial instruments: a passive system consisting of a large inertia mass and soft suspension for the higher frequency motion; an active system, electromagnetic or pneumatic, to drive a compliant test bed to counter the effects of long-period tilt. In this book we are concerned only with passive systems.

Transmissibility Ratio

The physical quantity that is taken as a measure of the objectionable consequence of site vibration is the relative displacement between the critical element and the seismic mass, as shown in Fig. 9.2b. Since the function of the mounting is to reduce this relative displacement to a small fraction of the corresponding site displacement, we express the performance of the mounting in terms of a relative displacement transmissibility ratio, which is defined as follows:

$$T_R = \frac{d_r}{d_s} \tag{9.1}$$

where T_R = relative displacement transmissibility ratio

d_r = maximum peak-to-peak (p–p) displacement of the critical element relative to the seismic mass (i.e., base, frame, bedplate of equipment) as shown in Fig. 9.2b

d_s = maximum p–p displacement of the support or site, as shown in Fig. 9.2a

Basis of Computations

The following is a brief outline of the method of computing T_R. The equations of motion of the two masses in Fig. 9.1c are written by applying Newton's laws of motion in much the same way as in the derivation of the force transmissibility in Chapter 8, but now the motion is initiated by the forces generated in the isolators when the support is displaced. Two simultaneous equations are obtained which involve the acceleration, velocity, and displacement of

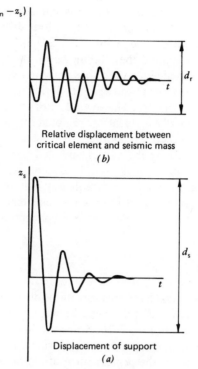

Figure 9.2. Definition of (*a*) displacement of support and (*b*) relative displacement of critical element.

both masses, the stiffness and damping coefficients of mounting and critical element, and the mass ratio. These are rewritten in terms of the following dimensionless ratios

$$\rho_m = f_m/f_s \qquad \text{mounting frequency ratio}$$
$$\rho_c = f_c/f_s \qquad \text{critical element frequency ratio}$$
$$\zeta_m = c_m/c_{cm} \qquad \text{mounting damping ratio}$$
$$\zeta_c = c_c/c_{cc} \qquad \text{critical element damping ratio}$$
$$R_c = m_c/m_m \qquad \text{critical element mass ratio}$$

where f_m = undamped natural frequency of mounting in its uncoupled vertical mode; that is, $m_m \, k_m$ system alone, assumed to be on a rigid support (Fig. 9.1*a*)

f_c = undamped natural frequency of critical element in uncoupled vertical mode; that is, $m_c k_c$ system alone, assumed to be on a rigid support (Fig. 9.1*b*)

f_s = damped frequency of damped harmonic, or frequency of sinusoidal excitation

c_{cm} = critical damping coefficient of mounting
c_{cc} = critical damping coefficient of critical element

Damped Harmonic Excitation

With transient excitation the equations of motion cannot be solved simply by assuming sinusoidal responses, as is done with sinusoidal excitation. They can be solved by mathematical methods (e.g., Waldersee, 1960) but today, with high speed digital computers readily available, it is convenient to use numerical integration. The displacement and velocity of each mass after a very short time interval are computed in terms of the initial conditions, by assuming that the motion can be described by a polynomial algebraic expression. The displacement and velocity after the first interval become the initial values for the next short interval and the process is repeated and continued until the "time histories" generated are of sufficient duration for the maximum positive and negative peak values of the relative displacement to be identified. From these the transmissibility ratio, as defined in Eq. (9.1), is determined. The details of the method are given by Macinante and Walter (1973, Appendix I), and the influence of integration step size and response duration are discussed by Macinante and Simmons (1975, Appendix I).

Sinusoidal Excitation

The rate of decay of the damped harmonic excitation becomes less rapid as the damping of the support or site decreases. By using the numerical integration procedure with the site damping ratio zero, the result obtained is that for sinusoidal excitation with the responses including the starting transients.

In the vibration literature, and in practice, the steady-state response is always understood to be that which persists after the starting transients have decayed. The data presented later for the steady-state transmissibility ratio are calculated from expressions derived by conventional mathematical analysis (see Macinante and Simmons, 1975, Appendix II). For the cases in which the mass ratio is negligible, the steady-state transmissibility ratio can be derived by treating the two-mass system as two one-mass systems in series. If also the system is completely undamped the following simple expression results (see Macinante and Simmons 1975, Appendix IV):

$$T_R = \frac{\rho_m^2}{(\rho_m^2 - 1)(\rho_c^2 - 1)} \qquad (9.2)$$

DESIGN DATA BASED ON TWO-MASS MODEL

The transmissibility ratio was computed for two values of the mass ratio of the critical element R_c: 0, 0.1. Three values were used for the damping ratio of the critical element ζ_c: 0, 0.1, 0.2, and for the mounting damping ratio ζ_m: 0, 0.1, 0.2. The values of natural frequency of the critical element and of the mounting are given later in the context of the particular series of computations.

The two values of the mass ratio and three values of each of the damping ratios form 18 combinations which, for convenience of reference, are identified by the "case numbers" given in Table 9.1.

In a first series of computations (Macinante and Walter, 1973) the transmissibility ratio was computed for all 18 cases, with excitation in the form of a particular damped harmonic which, for brevity, we refer to as the "datum" damped harmonic. Examination of the results identified those cases having the highest and the lowest T_R in a group with a common value of the mounting damping ratio. Two of these, one representing the worst-case undamped and the other the worst-case damped mounting with $\zeta_m = 0.2$, were investigated in more detail in a second series of computations using a wider range of damped harmonics, and sinusoidal excitation (Macinante and Simmons, 1975).

Response to "Datum" Damped Harmonic

The datum excitation is defined as a damped harmonic Fig. 9.2a having a predominant frequency of $f_s = 20$ Hz and damping ratio $\zeta_s = 0.2$. This is nominated as typical of the response of a ground or structural site to a shock or impulse occurring in the vicinity.

The critical element natural frequency values adopted were f_c: 10, 20, 30, . . . , 200 Hz, which cover the range mentioned in the introduction. The mounting natural frequency values used in the calculations were f_m: 1, 4, 7, . . . , 40 Hz. The values higher than 20 Hz, which is regarded as an upper limit in normal practice (Chapter 5), were included in order to overlap the excitation frequency of $f_s = 20$ Hz.

For each of the cases defined in Table 9.1, the transmissibility ratio was computed for all combinations of the mounting and critical element frequency

Table 9.1.   Case Numbers for Sensitive Equipment Installations

Mounting Damping Ratio ζ_m	Mass Ratio R_c	Critical Element Damping Ratio ζ_c		
		0	0.1	0.2
0	0^a	1	3	5
	0.1	2	4	6
0.1	0	7	9	11
	0.1	8	10	12
0.2	0	13	15	17
	0.1	14	16	18

[a] The condition $R_c = 0$ refers to a system in which the effective mass of the critical element is negligibly small in comparison with the seismic mass.

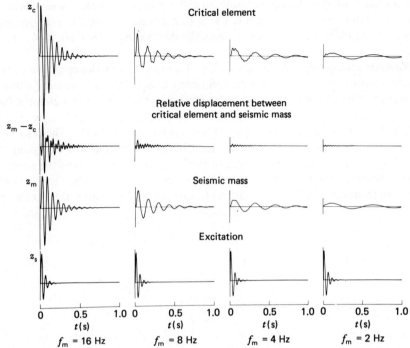

Figure 9.3. Examples of response to datum damped harmonic, showing how relative displacement decreases as mounting natural frequency has decreasing values: 16, 8, 4, and 2 Hz (Macinante and Walter, 1973).

ratios ρ_m and ρ_c respectively, formed by expressing the nominated values of f_m and f_c, as ratios of the excitation frequency $f_s = 20$ Hz. Thus,

$$\rho_m = f_m/f_s: \quad 0.05, 0.20, 0.35, \ldots, 2.0$$

$$\rho_c = f_c/f_s: \quad 0.5, 1.0, 1.5, \ldots, 10.0$$

Additional computations were made to determine T_R in the vicinity of resonance peaks where a small change in frequency ratio causes a large change in T_R.

The transmissibility ratio as defined depends on the peak–peak (p–p) value of the relative displacement as illustrated in Fig. 9.2*b* regardless of the shape (time history) of the response. Although the p–p value was determined by the computer without the need for computer plotting of the time histories, some were plotted as a matter of interest. For example, those in Fig. 9.3 show how the unwanted relative displacement is reduced by lowering the natural frequency of the mounting. The illustration refers to an equipment whose critical element has a natural frequency of 30 Hz, negligible damping, and mass ratio 0.1.

The four sets of time histories show how the relative displacement resulting from the datum damped harmonic excitation becomes progressively smaller as the mounting natural frequency takes the decreasing values 16, 8, 4, and 2 Hz.

Another interesting time history (Fig. 9.4) shows how the response of the critical element grows if its natural frequency is equal to the mounting natural frequency. The illustration refers to an equipment whose critical element has a natural frequency of 20 Hz, negligible damping, and negligible mass ratio, installed on an undamped mounting of natural frequency 20 Hz. The response is that resulting from the datum damped harmonic excitation. Referring to Fig. 9.4, the transient displacement of the support (trace a) starts the seismic mass vibrating (trace b) and, since the mounting is undamped, the seismic mass continues to vibrate at its natural frequency of 20 Hz after the excitation

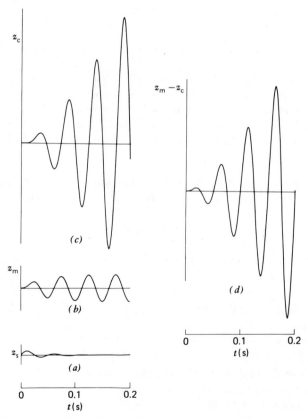

Figure 9.4. Resonance "build-up" of response of critical element in undamped system Case 1 (Table 9.1), when mounting, critical element, and excitation all have same frequency (Macinante and Walter, 1973).

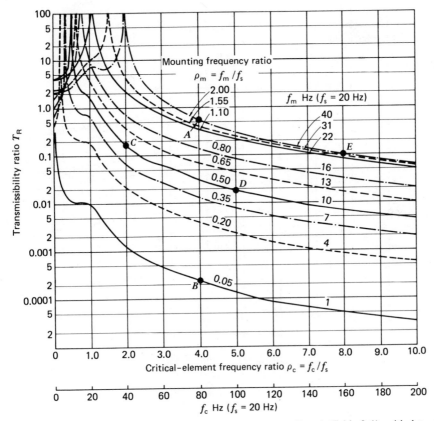

Figure 9.5. Transmissibility-ratio curves for undamped system Case 1 (Table 9.1), with datum damped harmonic excitation (Macinante and Walter, 1973).

has ceased. The critical element attached to the seismic mass has a natural frequency of 20 Hz, and since both its mass and damping are negligible its response builds up as shown in trace c. The final value attained by the rapidly growing relative displacement (trace d) is determined by the computation time allowed.

A set of transmissibility ratio (T_R) curves was plotted for each of the eighteen cases in Table 9.1, showing how T_R varies with the critical element frequency ratio for the various values of the mounting frequency ratio. The results for Cases 1 and 2, are illustrated in Figs. 9.5 and 9.6, respectively. For Case 1 each of the curves contains a sharp peak of high transmissibility where the critical element natural frequency equals the mounting natural frequency $(f_c = f_m)$. For Cases 2–18 the curves have reduced and rounded maxima. In the illustrations a bulge of increased T_R is evident where the critical element natural frequency equals the excitation frequency $(f_c = f_s)$.

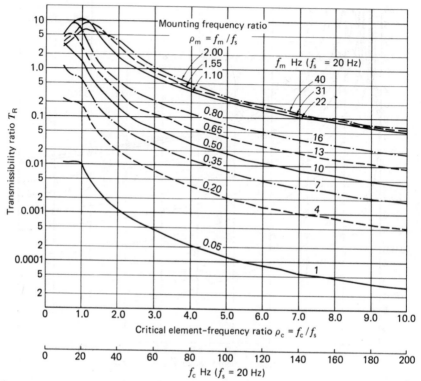

Figure 9.6. Transmissibility-ratio curves for undamped system Case 2 (Table 9.1), with datum damped harmonic excitation (Macinante and Walter, 1973).

An inspection of all 18 sets of transmissibility ratio curves revealed the following:

1. For a given value of critical element natural frequency in the region where $f_c > f_m$, the T_R decreases as the mounting natural frequency decreases. For example, assuming a Case 1 installation of an equipment having critical element natural frequency $f_c = 80$ Hz, it is seen in Fig. 9.5 that as the mounting natural frequency is decreased from 40 Hz (point A) to 1 Hz (point B) the T_R decreases from 0.6 to 0.0003.

2. For a given equipment on a given mounting, and with $f_c > f_m$, the transmissibility ratio decreases as the natural frequency of the critical element increases. For example, again referring to Fig. 9.5, assume that an equipment is installed on a mounting of natural frequency $c = 10$ Hz. If the critical element natural frequency f_c is increased from 40 Hz (point C) to 100 Hz (point D) the transmissibility ratio decreases from 0.15 to 0.02.

3. If the critical element natural frequency is high enough, a seismic mounting may be unnecessary. For example, consider a Case 1 installation of an equipment having a critical element natural frequency $f_c = 160$ Hz. The transmissibility ratio does not exceed 0.1 even if the mounting natural frequency is so high, (e.g., $f_m = 40$ Hz, point E), that the mounting is ineffective.

Identification of "Key" Cases

Comparisons of the results for the cases differing only in mass ratio showed T_R to be higher for $R_c = 0$ than for $R_c = 0.1$. Comparisons among cases differing only in critical element damping ratio showed T_R to be highest for $\zeta_c = 0$. Therefore, the worst case (i.e., highest T_R) in each group of six in Table 9.1 is the first (i.e., Cases 1, 7, and 13).

Comparison of the results for Cases 1, 7, and 13, which differ only in mounting damping ratio, showed that the T_R values for Case 7 were intermediate between the corresponding values for Case 1 and Case 13. Therefore, for the purpose of more detailed investigation Case 1 was adopted as the "key" case for an installation on an undamped mounting, and Case 13 the key case for an installation on a damped mounting ($\zeta_m = 0.2$).

Transmissibility Surfaces

For the two key cases, the way in which T_R is influenced by the two frequency ratios is shown in Figs. 9.7 and 9.8, which are computer-drawn surfaces with some hand-plotted contours added. The interesting feature in Fig. 9.7 is the ridge of high T_R on which any point represents an installation having equal values of the critical element and mounting natural frequencies. The reason for the high T_R is as follows. The motion of the seismic mass includes what would usually be called the starting transient, but in the present context it cannot properly be called a transient because the mounting is assumed to be undamped and therefore its free vibration theoretically goes on indefinitely and undiminished. This sinusoidal component of frequency f_m provides the excitation for the critical element, also undamped, which responds in resonance if its natural frequency is equal to the natural frequency of the mounting. Figure 9.7 shows, in three-dimensional form, a succession of resonance peaks some of which are shown in Fig. 9.5. For example, those for $f_c = f_m = 0.8$, 1.1, 1.55 and 2.0 in Fig. 9.5 are shown in Fig. 9.7 by the heavy broken lines. As mentioned in relation to Fig. 9.4, the maximum T_R attained under resonance conditions is limited by the computation time allowed; the maximum in Fig. 9.7 is 118.

The influence of damping of the mounting is seen in Fig. 9.8, which shows the T_R surface for Case 13. The diagonal region along which $f_c = f_m$ is now

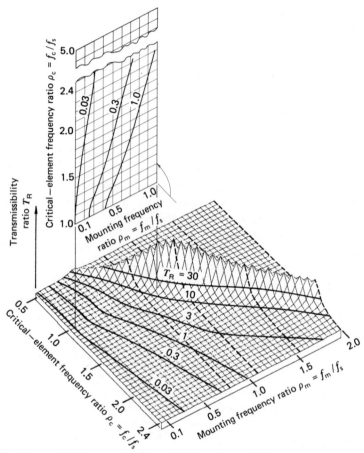

Figure 9.7. Transmissibility-ratio surface for undamped system Case 1 (Table 9.1), with datum damped harmonic excitation. Upper left diagram shows orientation of contour diagram in relation to surface. (After Macinante and Simmons, 1975).

relatively low and rounded. A bulge of high T_R is evident also in the region where f_c is in the vicinity of f_s (i.e., $\rho_c \approx 1$). The corresponding bulge is not evident in Fig. 9.7, which is scaled to show T_R of the order 100.

In designing a mounting for an equipment having a given critical element frequency ratio ρ_c, the designer will choose a mounting frequency ratio ρ_m such that the "design point," defined by coordinates ρ_c, ρ_m, is on or below a contour of a suitably low level. Therefore, we require design data in the form of T_R contours to facilitate selection of the design point.

Transmissibility Contours

Figures 9.9 and 9.10 illustrate the envelopes of the T_R contours for the groups of undamped and damped ($\zeta_m = 0.2$) mountings, respectively, in Table 9.1.

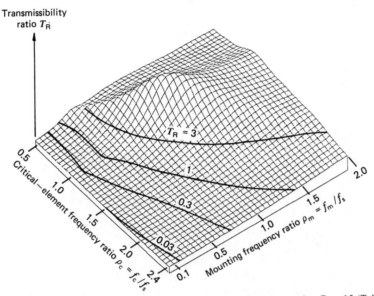

Figure 9.8. Transmissibility-ratio surface for system with damped mounting Case 13 (Table 9.1) and datum damped harmonic excitation (Macinante and Simmons, 1975).

Four transmissibility ratio levels are given, for $T_R = 0.03, 0.1, 0.3$, and 1.0, the lowest representing a very good mounting, the highest an ineffective one. The contours are identified only for the cases representing the highest and lowest T_R in the group. Thus, in Fig. 9.9, each crosshatched area is the envelope of the contours for Cases 1–6 for the particular T_R level. Similarly, Fig. 9.10 shows the envelopes of the contours for Cases 13–18. In the T_R-surface illustrations, the orientation is chosen to show, in the foreground, the region of low T_R that is of interest to the designer of the mounting. The orientation of the contour diagrams, in relation to the surfaces, is shown at the upper left in Fig. 9.7.

It may be noted in Figs. 9.9 and 9.10 that for a given critical element frequency ratio (e.g., $\rho_c = 2.2$ in Fig. 9.9) the T_R decreases as the mounting frequency ratio decreases (see points $A–D$). Also, for a given mounting frequency ratio (e.g., $\rho_m = 0.53$, Fig. 9.9) the T_R decreases as the critical element frequency ratio increases (see points $E–H$).

Example 9.1. As an example of the practical application of these data, assume that an equipment having a critical element frequency ratio of 2.2 is to be installed on an undamped mounting. A T_R of 0.03 is assured if the mounting frequency ratio is 0.25 (point D, Fig. 9.9), and T_R of 0.1 is assured if the mounting frequency ratio is 0.45 (point C).

The left boundary of the T_R envelope is the contour for the worst-case (highest T_R) in each group; this is the contour for an equipment having an

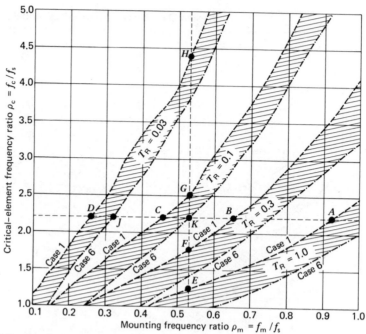

Figure 9.9. Envelopes of contours of four T_R levels for systems with undamped mounting (Cases 1 through 6, Table 9.1), with datum damped harmonic excitation (Macinante and Walter, 1973).

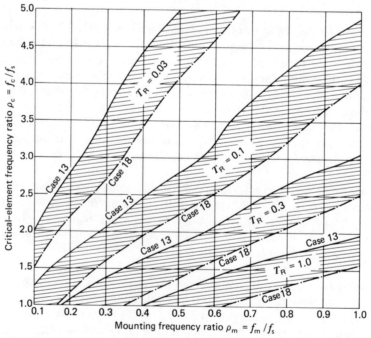

Figure 9.10. Envelopes of contours of four T_R levels for systems with damped mounting (Cases 13 through 18, Table 9.1), with datum damped harmonic excitation (Macinante and Walter, 1973).

undamped critical element of negligible mass ratio. It follows that mounting design based on the left boundary is conservative (i.e., the T_R will be less than the nominated value) if the damping ratio and mass ratio of the critical element are not negligibly small. For example, if the critical element damping ratio is $\zeta_c = 0.2$, the mass ratio $R_c = 0.1$, and the mounting is designed on the "worst-case" basis, the mounting frequency ratio chosen to assure $T_R = 0.1$ is $\rho_m = 0.45$ (point C). In fact the installation is Case 6 (Table 9.1), for which $\rho_m = 0.53$ (point K) would give $T_R = 0.1$. The lower value $\rho_m = 0.45$ (point C) provides a T_R somewhat lower than 0.1 as indicated by the position of C between points J and K on the $T_R = 0.03$ and 0.1 contours for Case 6.

Response to Wider Range of Damped Harmonics

For the two worst Cases 1 and 13, the transmissibility ratio was computed for the wider range of damped harmonics defined by $f_s = 20$ Hz; $\zeta_s = 0.3, 0.2, 0.1, 0.01$, which are illustrated in Fig. 9.11. The results for the condition

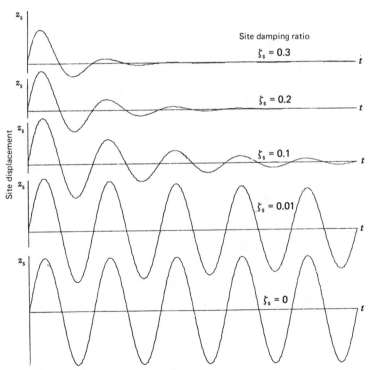

Figure 9.11.   Range of excitation used in the computations.

$\zeta_s = 0$, which of course represents a sinusoidal excitation, are given in the next section. The frequency ratios used in the computations were

$$\rho_m = f_m/f_s: \quad 0.1, 0.2, 0.3, \ldots, 1.0$$
$$\rho_c = f_c/f_s: \quad 1.0, 1.5, 2.0, \ldots, 5.0$$

Supplementary computations to determine the influence of integration step size and response duration showed that the T_R calculated for any pair of frequency ratio values with f_s: 10, 40 Hz did not differ by more than one percent from the value obtained with $f_s = 20$ Hz (see Macinante and Simmons, 1975, Appx. I). Therefore, for practical design purposes the data are applicable for excitation frequency in the range f_s: 10–40 Hz, and mounting and critical element natural frequencies in ranges corresponding to the frequency ratios ρ_m: 0.1–1.0; ρ_c: 1.0–5.0. Thus for $f_s = 10$ Hz, the ranges are f_m: 1–10 Hz, f_c: 10–50 Hz; and for $f_s = 40$ Hz, the ranges are f_m: 4–40 Hz, f_c: 40–200 Hz.

For each of the damped harmonics a set of contours of the four T_R levels was derived. The results for the undamped and damped key cases are given in Figs. 9.12 and 9.13, respectively. Each crosshatched area is the envelope

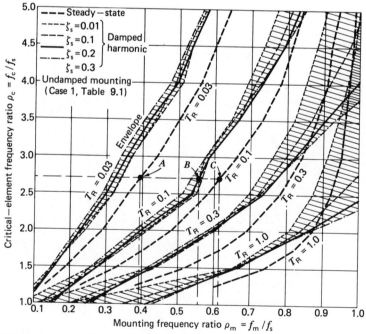

Figure 9.12. Contour envelopes for undamped system Case 1 (Table 9.1). Each envelope contains contours for four damped harmonic excitations (Fig. 9.11) for the T_R level shown (Macinante and Simmons, 1975).

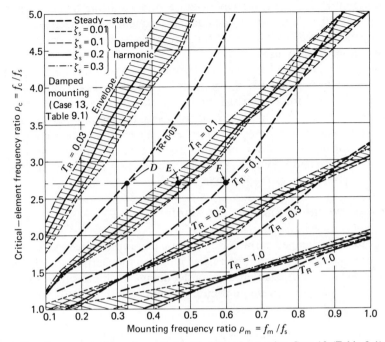

Figure 9.13. Contour envelopes for system with damped mounting Case 13 (Table 9.1). Each envelope contains contours for four damped harmonic excitations (Fig. 9.11) for the T_R level shown (Macinante and Simmons, 1975).

of the four contours for ζ_s: 0.3, 0.2, 0.1, 0.01, for the T_R level indicated. The heavy full line in each area is the contour for the datum excitation $\zeta_s = 0.2$. The four heavy broken lines relate to the steady-state condition, which is discussed later.

By using the contour of the desired T_R level for a particular damped harmonic excitation as a relationship between critical element and mounting frequency ratios, the value can be found for the mounting frequency ratio that is required to achieve a nominated T_R for that particular excitation. Or, by using the left boundary of the envelope of the contours of a nominated T_R level, the value indicated for the mounting frequency ratio is such that the nominated T_R is attained for any damped harmonic having site damping ratio in the range ζ_s: 0.01–0.3 (see Example 9.2).

Response to Sinusoidal Excitation

When a sinusoidal excitation "starts up," the mounting and the critical element respond with their natural vibrations at the frequencies f_m and f_c, respectively, which are superposed on their responses at the forcing frequency f_s. Damping of the mounting and critical element causes these starting transients to decay, leaving only the steady-state response at frequency f_s. The transmissibility ratio was determined for both conditions: response including starting transients, and steady-state response.

Results are presented for the key (worst) cases of undamped and damped mountings, Cases 1 and 13, respectively. For both cases the critical element is undamped, and for Case 1 the mounting also is undamped. This involves the slight difficulty in terminology mentioned in the discussion relating to Fig. 9.7, that the natural responses of critical element and mounting cannot properly be called transients if there is no damping to cause their decay. Nevertheless, for convenience and brevity, we discuss the cases involving these undamped natural responses under the following heading.

Response Including Starting Transients

The response to damped harmonic excitation that is computed by the numerical integration procedure described earlier includes the starting transients of the mounting and the critical element. As the damping ratio decreases the rate of decay of the damped harmonic diminishes until, with $\zeta_s = 0$, the excitation

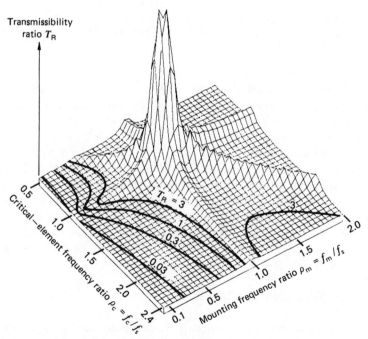

Figure 9.14. Transmissibility-ratio surface for undamped system Case 1 (Table 9.1), for sinusoidal excitation with response including starting transients (Macinante and Simmons, 1975).

is sinusoidal (see Fig. 9.11). The results for this condition are referred to as those for sinusoidal excitation with the response including the starting transients.

Undamped Mountings.   The transmissibility ratio surface for the key case undamped system, shown in Fig. 9.14, now has three ridges of high T_R with a very high peak where they intersect. For this completely undamped system theoretically the T_R tends to infinity along these ridges. The variation in height along a ridge indicates variation in the rate of growth of T_R. The development of these ridges can be explained in the following way by considering the two-mass system to behave as two one-mass systems in series. This is permissible for a system with $R_c = 0$ because the seismic mass is "unloaded" by the critical element of negligible mass.

The vibration of the seismic mass involves two component frequencies: the forcing frequency f_s and the frequency f_m of the undamped starting "transient." These two components excite the critical element in the following ways.

First, consider the region in Fig. 9.14 where $\rho_c > 1$. An equipment having $\rho_c = 1.5$ (Fig. 9.15), for example, can experience high T_R if the mounting frequency ratio is ρ_m: 1.0 or 1.5, as shown by the two peaks in the profile $A'B'$. The first peak, which is on the ridge CD results from the resonance of

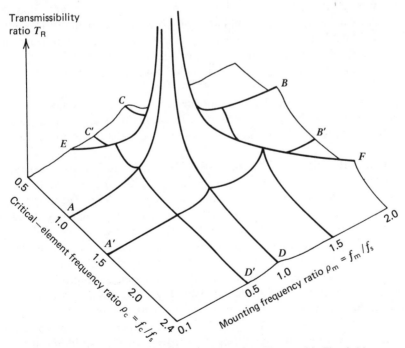

Figure 9.15.   Resonances associated with surface illustrated in Fig. 9.14.

a mounting having $\rho_m = 1$ ($f_m = f_s$). This resonance response of the seismic mass provides the excitation of the critical element. The resulting T_R, which is determined by the relative displacement of critical element and seismic mass, increases slowly for high values of the critical element frequency ratio (ρ_c near point D) and more rapidly as ρ_c approaches unity where the critical element also is in resonance.

The second peak on the line $A'B'$, which is on the ridge EF, results from the fact that when $\rho_m = \rho_c = 1.5$ the mounting natural frequency is equal to the critical element natural frequency ($f_m = f_c = 1.5 f_s$). As discussed in relation to Fig. 9.7, the critical element is excited in resonance by the starting "transient" of the undamped mounting, which is a sinusoidal vibration of frequency $1.5 f_s$.

The ridge AB indicates the high T_R that can occur if the critical element natural frequency is equal to the excitation frequency ($\rho_c = 1$). On the part of this ridge in the vicinity of point A the T_R builds up slowly because the response of the seismic mass at the forcing frequency f_s, which provides the excitation of the critical element, is relatively small where the mounting frequency ratio is low. The profile of this ridge illustrates how much more rapidly the T_R increases as the mounting frequency ratio approaches unity.

Although the surface in Fig. 9.14 may seem to be of only academic interest because the hardware of actual mountings and critical elements cannot have zero damping, it is also of practical value as a graphic reminder that if an equipment whose critical element has negligible mass ratio and little damping is installed on a lightly damped mounting, it can experience high transmissibily as a result of a lightly damped transient excitation, or the starting of a source of steady-state excitation, if any pair or all three frequencies f_c, f_m, and f_s happen to have the same value.

In the region well away from resonance ($T_R < 1$), the T_R contours for Case 1, for sinusoidal excitation with response including the starting transients, did not differ significantly from those identified for $\zeta_s = 0.01$ in Fig. 9.12.

Damped Mountings. The transmissibility ratio surface for the key case damped mounting, Case 13, with response including the starting transients, is shown in Fig. 9.16. Now there is only one region of high T_R, which results from resonance of the critical element when its natural frequency equals the excitation frequency. The response of the critical element is reduced because damping of the mounting limits motion of the seismic mass at the forcing frequency f_s which provides the excitation of the critical element.

The other two ridges that develop with an undamped mounting (Fig. 9.14) are now suppressed because the damping quickly attenuates the starting transient of the mounting at its natural frequency f_m. Consequently, the resonances involving the coincidences $f_m = f_s$ and $f_m = f_c$ do not develop. The T_R contours for Case 13 with response including the starting transients did not differ significantly from those identified for $\zeta_s = 0.01$ in Fig. 9.13.

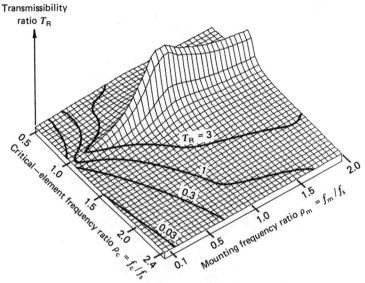

Figure 9.16. Transmissibility-ratio surface for system with damped mounting Case 13 (Table 9.1), for sinusoidal excitation with response including starting transients (Macinante and Simmons, 1975).

Steady-State Response

The T_R for the response not including the starting transients, usually called the steady-state response, was computed for all 18 cases. The results were examined so that key cases for steady-state response could be identified independently, not assumed to be the same as those for damped harmonic excitation. It was found that the key cases are again Cases 1 and 13 (Macinante and Simmons, 1975, Appx. III).

Undamped Mountings.   The T_R surface for the steady-state response for the key case undamped mounting is given in Fig. 9.17. There are two sharp ridges of high T_R resulting from resonances of the critical element ($f_c = f_s$), and of the mounting ($f_m = f_s$). There is no diagonal ridge representing coincidence of f_c and f_m, because in the steady-state condition the system vibrates at the one frequency f_s; there is no vibration of the seismic mass at the mounting natural frequency f_m to excite the critical element.

The T_R contours for this case are included in Fig. 9.12, in which the heavy broken lines are the contours for the four indicated levels. In the region where $T_R < 0.1$ the contours for the steady-state condition are to the right of the corresponding envelope of the contours for damped harmonic excitation. This means that for given equipment (ρ_c fixed) to satisfy a specified requirement (e.g., $T_R = 0.1$) when installed on an undamped mounting, a higher mounting

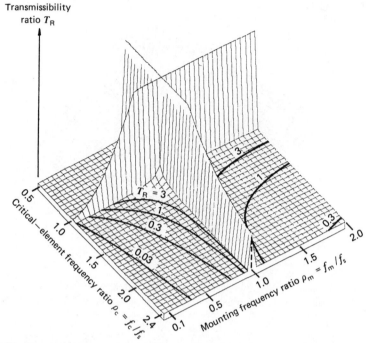

Figure 9.17. Transmissibility-ratio surface for steady-state response of undamped system Case 1 (Table 9.1). (Macinante and Simmons, 1975).

frequency ratio (stiffer mounting) is acceptable for a steady-state excitation (e.g., point C) than for a damped harmonic of the same frequency (e.g., point B).

Or, in terms of transmissibility ratio, a mounting designed to have a specified T_R when subjected to damped harmonic excitation will have a somewhat lower T_R when subjected to steady-state vibration of the same frequency. For example, consider an installation that requires a mounting frequency ratio of $\rho_m = 0.55$ (point B, Fig. 9.12) to achieve $T_R = 0.1$ when subjected to the datum damped harmonic excitation. The same installation under steady-state conditions has the lower T_R indicated by the position of point B between points A and C on the 0.03 and 0.1 steady-state contours.

Damped Mountings. The transmissibility ratio surface for the steady-state response of the key case damped mounting, Case 13, shown in Fig. 9.18, has only one ridge, any point on which represents resonance of a critical element whose natural frequency is equal to the excitation frequency ($f_c = f_s$). The ridge that represents resonance of the mounting in Figs. 9.14 and 9.17 does not develop with a damped mounting, because the damping limits the amplitude of the displacement of the seismic mass at the forcing frequency f_s to a finite

Transmissibility
ratio T_R

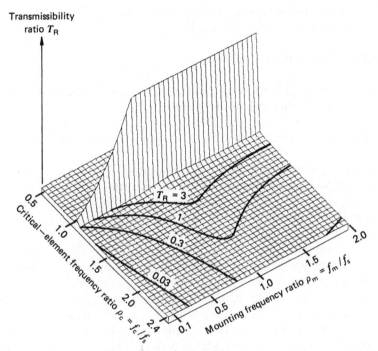

Figure 9.18. Transmissibility-ratio surface for steady-state response of system with damped mounting Case 13 (Table 9.1). (Macinante and Simmons, 1975).

value, and the response of the critical element also is finite except where the critical element natural frequency equals the excitation frequency.

The steady-state contours are shown in Fig. 9.13 as the heavy broken lines for the four nominated T_R levels. In the region where $f_m/f_s < 0.8$ these contours are to the right of the corresponding envelopes of the T_R contours for damped harmonic excitation. In this region a damped mounting designed to have a specified T_R for damped harmonic excitation (e.g., point E) will have a somewhat lower T_R for steady-state excitation of the same frequency. This can be seen by noting the position of the point E between points D and F on the 0.03 and the 0.1 steady-state contours.

SUMMARY OF RESPONSE OF TWO-MASS MODEL

We now briefly review the main features of the behavior of the two-mass model of a sensitive-equipment installation subjected to site vibration. We note those relationships among the design variables which should be avoided because they lead to high transmissibility ratio, and those which should be designed into the installation to achieve a low transmissibility ratio. In particular we show how the quality of vibration isolation achieved by a mounting is influenced by the characteristics of the critical element.

Resonance Conditions

Resonance conditions are best described with reference to the installation most likely to respond with troublesome resonances. This is an equipment having a lightly damped critical element of negligible mass ratio, installed on a lightly damped mounting, which is idealized as a completely undamped system (Case 1, Table 9.1).

When the site of such an installation is subjected to a lightly damped harmonic or a sinusoidal vibration, a resonance of the mounting occurs if the mounting natural frequency f_m is equal to the excitation frequency f_s. The effect of this is evident in the ridge of high T_R in the region of $\rho_m = 1$ in Figs. 9.14 and 9.17.

Usually these two frequencies are unequal, and the vibration of the seismic mass has two components: the starting transient, which is a lightly damped and hence quasi-sinusoidal vibration of frequency f_m; and the sinusoidal forced vibration at frequency f_s. This motion of the seismic mass provides the excitation of the critical element, with the possibility of resonance if the critical element natural frequency f_c is equal to the mounting natural frequency f_m. The effect of this is indicated by the diagonal ridge of high T_R in Fig. 9.14. The development of this resonance does not require a sustained site vibration. A transient of short duration, for example, the datum damped harmonic, can start the seismic mass vibrating at its natural frequency and, since the mounting is lightly damped, the seismic mass continues to vibrate at its natural frequency f_m. If the natural frequency of the critical element is equal to that of the mounting, the resonance response builds up as described in relation to Fig. 9.7. Damping of the mounting causes the vibration of the seismic mass to decay more rapidly, reducing the duration of the excitation of the critical element, with the result that the diagonal region is much lower, as in Fig. 9.8.

A third possible resonance is that of an equipment whose critical element natural frequency f_c is equal to the frequency f_s of the forced component of vibration of the seismic mass. The effect of this is indicated by the zone of high T_R in the region of $\rho_c = 1$ in Figs. 9.14–9.18. Figure 9.16 shows the influence of damping of the mounting which reduces the forced vibration of the seismic mass at the forcing frequency f_s, and hence reduces the level of excitation of the critical element.

Design Area

Design to avoid the effects of resonance is an obvious and essential requirement, which can be satisfied by ensuring that there is no coincidence of any pair or all three of the frequencies f_s, f_m, and f_c. By avoiding any such coincidence the design point is placed in one of the six regions of relatively low T_R bounded by the ridges in Fig. 9.14.

However, only a limited area in these regions has T_R low enough for effective vibration isolation. The design point must be in the "design area" where the T_R is well below unity. This is the area, along the lower left side of the various T_R surfaces illustrated, that is bounded by the contour of the nominated maximum T_R; for example, $T_R = 0.3$ in Fig. 9.19.

Within the design area, details of the relationship of T_R to the two frequency ratios ρ_m and ρ_c are condensed in Figs. 9.12 and 9.13 by using envelopes of the T_R contours to minimize the information required about the equipment and the excitation. By using the left boundary of a contour envelope as the design relationship, the design is conservative as illustrated in Example 9.1.

The essential data for design for $T_R < 0.1$ on this basis are further condensed in Fig. 9.20. The broken lines are the left boundaries of the T_R: 0.03, 0.1 envelopes in Fig. 9.12 (undamped mountings), and the two full lines are the left boundaries of the T_R: 0.03, 0.1 envelopes in Fig. 9.13 (damped mountings, $\zeta_m = 0.2$).

Interpolation between specific computed values of T_R is permissible in areas remote from the sharp ridges of high T_R caused by the responses of undamped

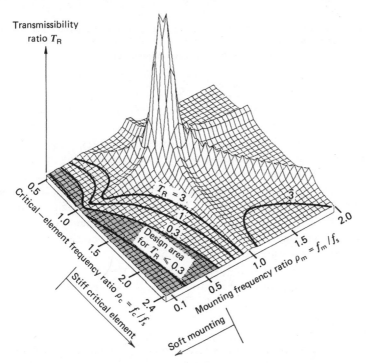

Figure 9.19. "Design area," "stiff critical element," and "soft mounting" areas on transmissibility-ratio surface.

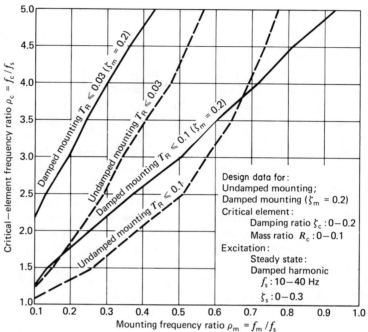

Figure 9.20.   Design data for undamped and damped systems (Case 1 and Case 13, Table 9.1), showing mounting frequency ratio required to assure T_R: 0.03, 0.1, with excitation and installation parameters in the ranges shown. (Macinante and Simmons, 1975).

elements. In support of this assertion we refer to the physical nature of the system and the topography of the T_R surfaces which were generated in the computer with small intervals in ρ_m (0.05) and ρ_c (0.05).

A mounting having the natural frequency derived by using the left-boundary lines as the relationship between critical element and mounting frequency ratios will ensure that the nominated T_R will not be exceeded for any damped harmonic excitation of frequency in the range f_s: 10–40 Hz and site damping ratio in the range ζ_s: 0.01–0.3 and for any steady-state excitation, for equipment and mounting parameters in the ranges covered by the computations as detailed under Design Data Based on Two-Mass Model.

Within the design area, and within the stated ranges of the design variables and excitations, we can make the following generalizations.

The Mounting

For a given equipment and excitation frequency, which together determine the critical element frequency ratio, the T_R decreases as the natural frequency of the mounting decreases.

For a given mounting natural frequency, the T_R is lower for an undamped than a damped mounting. However, for practical reasons, damping may be necessary, even though it raises the T_R. For example, the simplest form of undamped mounting is that using helical spring isolators, but these are unsuitable for extremely sensitive equipment because of transmission resulting from internal vibrations among the coils. Also damping of some mountings is necessary to restore the equipment quickly to its level position after manipulation or adjustment.

The Critical Element

For given mounting and excitation parameters, the T_R decreases as the natural frequency of the critical element increases. This is illustrated in the discussion relating to Fig. 9.5 which shows that even with an ineffective mounting (mounting frequency ratio $\rho_m = 2.0$) the T_R is below 0.1 if the critical element frequency ratio is high enough (e.g., $\rho_c > 8$, see Fig. 9.5, point E). The vibration of the critical element relative to the seismic mass is very small because the critical element is very stiff.

Damping of the critical element is advantageous: the T_R decreases with increase of the damping, other factors remaining unchanged.

Risk Involved in Ignoring Critical Element

Use of the one-mass model assumes that the only resonance possible is that of the whole equipment on the mounting. In practice this resonance is avoided by designing for a low value of the mounting frequency ratio. This places the design point in the region marked "soft mounting" in Fig. 9.19. If, unluckily, the critical element natural frequency is close to the site vibration frequency, the design point will fall in a region of relatively higher T_R in the vicinity of $\rho_c = 1$. This can be seen in Fig. 9.19 and in the illustrations of the various T_R surfaces.

There is also the possibility of high T_R if the critical element natural frequency is equal to the mounting natural frequency, for this places the design point on the diagonal ridge, Figs. 9.7 and 9.14. However, if, as is normal practice, the mounting frequency ratio is suitably low (e.g., $\rho_m < 0.3$) trouble from coincidence of f_c and f_m can occur only if the critical-element frequency ratio is exceptionally low.

PRACTICAL APPLICATION OF THE DESIGN DATA

In the usual design situation a mounting is required to isolate a particular item of equipment from existing or anticipated conditions of site vibration, and the data are used to determine the value that should be specified for the natural frequency of the mounting to ensure a specified quality of vibration isolation. The designer's task then is to realize this natural frequency in the hardware by selecting isolators of a suitable type and stiffness, and arranging their layout, using the methods given in Chapter 7.

If the equipment to be installed has a high critical element frequency ratio the data may show that a seismic mounting is unnecessary. This is indicated if the required transmissibility ratio is attainable with a mounting natural frequency of, say, 20 Hz or higher, in which case the mounting would be no "softer" than most sites.

The data may be used also by designers of sensitive equipment to determine the natural frequency that should be "designed into" the critical element to ensure that its response will be within specified limits when the equipment is installed on a mounting of specified frequency ratio, and subjected to site vibration within specified limits of frequency and amplitude.

In order to use the data in the design of a mounting, some information is required about the site vibration and the characteristics of the equipment, and a decision must be made about the desired quality of isolation. These matters are now discussed.

Determine Site Vibration Characteristics

The frequencies and amplitudes of the major components of the site vibration in the range up to say 100 Hz, should be determined, using, if necessary, the measurement services of vibration consultants. Measurements should be made at selected points or locations at times and under conditions that represent the most severe excitation.

Preliminary measurements should be made in the vertical and two horizontal directions. Then, if the excitation is obviously more severe or significant in a particular direction, more detailed measurements can be made for that direction (e.g., Example 3.6). The vertical vibration of a suspended floor is usually more troublesome than the horizontal. At the upper levels in a building, in addition to the vertical vibration of the suspended floors the horizontal vibration of the whole building in its flexural modes may be significant. This is particularly important in a tall building because the horizontal vibration frequency in the lower flexural modes is typically well below 10 Hz (e.g., Example 2.3) and this motion is difficult to isolate with a passive mounting. It is worth repeating (see Chapter 3, Siting of Sensitive Equipment) that sensitive equipment should not be installed at upper levels in tall buildings because of the possibility of trouble with horizontal vibration, and because floors tend to be of lighter construction and hence "lively" at the upper levels.

Although the mounting natural frequency is usually determined in relation to the lowest excitation frequency, any higher-frequency components should be identified so that the mounting can be designed to avoid coincidence of a natural frequency in any mode with an excitation frequency.

The site vibration may include ground or structural responses to shock sources. If there is a major shock source such as a forging hammer, a record of the transient response can be made, from which the predominant frequency and site damping ratio can be found (Chapter 2, Transient Vibration). However, the difficulty arises here that the transient response frequency observed at a nominated site may be significantly different from the response frequency when the site is loaded by the installation. It is acknowledged in Chapter 6 in the section Case for Adoption of Two-Mass Model, Excitation at Support of Sensitive Equipment, that the response of the installation is influenced by characteristics of the site that are not represented in the two-mass model.* In using the design data derived from the two-mass model, the designer needs to make an estimate of the predominant frequency f_s of the transient response of the site when loaded by the installation. The estimation of the transient response frequency of the loaded site on the basis of measurements made at the unloaded site is a matter for the soil dynamics or the structural dynamics specialist [e.g., see Barkan (1962); Biggs (1964)]. With the procedure given below, it is a simple matter to check the mounting design for site vibration frequencies in a range covering the uncertainty or inaccuracy of the estimated value.

In the design of certain installations, for example, facilities for testing inertial guidance accelerometers mentioned earlier (Gray et al., 1972), sensitive equipment must be isolated from the low-frequency, long-period disturbances inherent in the earth. These are too low in frequency (a small fraction of 1 Hz) to be isolated by a passive seismic mounting, which would need to have a natural frequency significantly lower than that of the disturbance. The method presented in this chapter is applicable only to the design of passive mountings having natural frequency down to about 1 Hz, and hence capable of isolating frequencies of a few hertz and higher.

Determine Critical-Element Characteristics

If the critical element is obvious and accessible, its dynamic characteristics can be found by the simple experimental methods mentioned below. Unfortunately, in some equipment, for example, an electron microscope, the critical element may be inaccessible and unidentifiable by the designer of the mounting. The information required about the nature and characteristics of the critical element is known or could be determined by the designer of the equipment, but is rarely available to the designer of the mounting. It is hoped that a wider realization of the importance of the natural frequency and damping of the

* See also discussion by G. Harold Klein [*Shock and Vibration Digest*, **8**(11), 36 (1976)] on the author's paper [*Shock and Vibration Digest*, **8**(7), 3–24 (1976)], and author's reply [*Shock and Vibration Digest*, **9**(5), 4 (1977)].

critical element will stimulate the suppliers of vibration-sensitive equipment to make this information available.

Natural Frequency

For a critical element which is a simple structural component, and whose mass ratio is low, the effects of coupling can be ignored and the natural frequency determined from the response to an impact applied manually or with a soft-faced hammer. The response can be detected with a small accelerometer attached to the critical element, or with a proximity transducer arranged to detect the response relative to an adjacent part of the equipment (e.g., see Example 5.12). The frequency of a transient measured in this way is the "damped natural frequency," which differs from the undamped natural frequency (defined as f_c in the design data) to an extent that depends on the damping. The difference is negligible with damping ratio below about 0.2 (see Den Hartog, 1956, p. 40).

The natural frequency of the critical element will vary to some extent with the configuration of the equipment, for example, the height and radial position of an adjustable element such as an optical head, or the cutting head of a machine tool, and with the condition of adjustment of slides and the tightness of joints. The natural frequency should be determined for the mode that is important to the functioning of the equipment, even if this is not the mode represented in the design model. The question of the "fit" or relevance of the model to the actual installation (Chapter 6) is a matter for separate consideration.

The natural frequency of the critical element is determined experimentally with the equipment idle and the critical element not in contact with an adjacent part of the equipment. A defense of the "no-contact" assumption in the application of the two-mass model to machine tool installations is given above under Design Model. For sensitive equipment in general, there may be contact through a probe or stylus between critical element and base, or between two critical elements. Generally, this would involve small contact forces having negligible influence on the response of the system.

If experimental determination is not practicable, the natural frequency may be estimated by calculation. A great deal of basic information is available in the technical literature on the vibration of structural members such as cantilevers, beams, shafts and frames; and techniques are well established for the calculation of the natural frequencies of structural and mechanical elements of various geometric forms and material properties (e.g., see Harris and Crede 1976, Chap. 1, Appx. 1.1; Steinberg, 1973). In assigning a design value for the natural frequency it must be realized that the calculated value is likely to be higher than the actual value if the calculation involves the assumption of clamped (rigidly fixed) ends. The end fixing of actual structural elements may be much more compliant because of the influence of joints, bearings and guideways.

Damping Ratio

If the natural frequency of the critical element is determined experimentally from a recorded free vibration transient, the damping ratio can be found from the same record (Example 2.2). Experimental determination of the damping ratio is not essential, because the transmissibility ratio decreases with increasing damping of the critical element, and the design procedure is conservatively based on the assumption of zero damping.

Mass Ratio

The design data have been computed for installations with mass ratio 0 and 0.1. Since the design is conservatively based on the assumption that the mass ratio is close to zero, there is no need to calculate the effective mass of the critical element if it is obviously less than one-tenth of the total seismic mass.

Nominate Transmissibility Ratio

In some applications a numerical value can be assigned to the permissible relative displacement of the critical element. If also the site vibratory displacement is known, a realistic transmissibility ratio can be specified.

For example, a specified quality of the surface finish produced by a grinding machine may determine the permissible vibratory displacement of the wheel relative to the workpiece. In observations with an optical interferometer, fringes that should be sharp may be blurred as a result of site vibration. For example, in a certain application of multiple-beam interferometry, blurring of one-tenth of a fringe spacing is objectionable. This corresponds to a relative displacement of the optical returning surfaces of 0.025 μm. If the interferometer is sited in a laboratory where the floor vibratory displacement is of the order 0.1 μm the maximum permissible transmissibility ratio is $0.025/0.1 = 0.25$.

The quality of isolation specified should be such that the equipment can be operated at its maximum rated capability (e.g., highest quality of surface finish for a machine tool, maximum sensitivity of a measuring instrument, highest magnification of optical equipment or electron microscope), even though current work contracts may not require the maximum capability.

In many applications the aim is simply to make T_R as low as possible within the constraints of cost and any other considerations involved. A special mounting is hardly worthwhile if it does not attain T_R less than say 0.3. On the other hand, there is no point in nominating a design value so low that the displacement of the critical element resulting from external vibration would be less than that generated by internal sources such as motors, pumps, and other accessories, essential to the operation of the equipment. If there is no functional or performance criterion on which T_R may be based, a value of 0.1 is suggested as a reasonable design objective.

Assign Damping Ratio of Mounting

Within the design area, theoretically an undamped is preferable to a damped mounting in that, for a given equipment subjected to given conditions of excitation, the T_R is higher for the damped mounting. This is illustrated later in Example 9.2. Nevertheless for some installations damping must be provided and the somewhat higher T_R accepted.

For example, an instrument such as a microbalance must be at rest and in its level (horizontal) position during measurements. An undamped mounting is objectionable because the instrument takes a long time to return to its level position following adjustment or manipulation by the operator. Mountings for such instruments require some means of self-leveling, or sufficient damping to suppress the free vibration quickly. Another type of installation in which an undamped mounting may be ineffective is that of an extremely sensitive instrument such as an optical interferometer installed on undamped isolators in the form of helical springs, which may be disturbed by vibration transmitted around the coils of the springs.

The damping ratio of isolators of the kinds described in Chapter 5 can be expected to lie in the range ζ_m:0–0.2. Since the actual value cannot be determined until the installation is completed, the damping ratio assumed for design purposes must be based on experience and/or data supplied by isolator manufacturers. The design data are detailed for the values $\zeta_m = 0$ and 0.2. For isolators having damping ratio in the vicinity of 0.1 the data can be interpolated between 0 and 0.2, or the data for $\zeta_m = 0.2$ used, in which case the design is conservative.

Evaluate Mounting Natural Frequency

From the values assigned to the critical element natural frequency f_c and the site vibration frequency f_s, evaluate the critical element frequency ratio $\rho_c = f_c/f_s$. Refer to the appropriate line in Fig. 9.12 or 9.13 and read the value of mounting frequency ratio ρ_m corresponding to the value of the critical element frequency ratio ρ_c and the nominated transmissibility ratio. The required mounting natural frequency is then evaluated as $f_m = \rho_m f_s$.

Before illustrating this procedure with a numerical example we should discuss the question of what is the "appropriate" line in Figs. 9.12 and 9.13 to use as the design relationship, because T_R varies not only with the two frequency ratios ρ_c and ρ_m, but also with the mass ratio and the damping ratios of critical element and mounting (the 18 cases in Table 9.1), and with the form of excitation (the five conditions illustrated in Fig. 9.11). A complete design data for all these cases and conditions would comprise 90 sets of the contours of the four nominated levels of T_R: 0.03, 0.1, 0.3, 1.0.

For convenience of presentation and practical application, these data have been condensed in the manner detailed under Design Data Based on Two-Mass Model. Briefly, the condensation involves the following.

1. Ignoring differences in critical element damping ratio and mass ratio, within the limits ζ_c: 0–0.2, and R_c: 0–0.1, and basing the design on the "worst case," that having the highest T_R in a group having a common mounting damping ratio. This is done by using, as the design relationship between the two frequency ratios ρ_c and ρ_m, the left boundary of the envelope of the relevant group of contours, as explained in relation to Figs. 9.9 and 9.10.

2. Ignoring differences in the excitation within the range shown in Fig. 9.11, and basing the design on the worst case, that with the highest T_R in the group of five excitations. This is done by using as the design relationship the left boundary of the envelope of the contours relating to the five forms of excitation as explained in relation to Figs. 9.12 and 9.13.

The designer of the mounting may use the data in the less condensed and less conservative form if the necessary information is available about the equipment and the excitation, as indicated in Example 9.2.

Select Isolators and Arrange Layout

The procedure for selecting and arranging the isolators to achieve the required value of the natural frequency is the same as that given for machinery mountings in Chapter 8.

Example 9.2. A seismic mounting is required for a surface grinder which will be subjected to site vibration from two major sources: a forging hammer and a Diesel set (a standby Diesel-electric generator). The maximum permissible vibratory displacement of the grinding-wheel head resulting from these external sources of vibration, during finishing cuts, is specified as 3 μm (p-p) vertical, relative to the horizontal surface being ground. The measured characteristics of the site vibration and of the wheel head of the surface grinder are given below. Determine the value that should be specified for the vertical natural frequency of the mounting, considering the alternatives of helical spring and rubber isolators.

Operation of the forging hammer produces at the site of the proposed installation a transient vibration that is approximately a damped harmonic having a predominant frequency of 18 Hz, damping ratio 0.1, and vertical displacement 40 μm p-p.

The Diesel set produces a sinusoidal vibration at the shaft rotational frequency of 1000 rev/min (16.7 Hz). The vertical vibratory displacement at the surface grinder site is 20 μm rms.

From a record of the vertical free vibration of the wheel head of the surface grinder, the natural frequency is approximately 38 Hz and the damping ratio 0.1.

In the nomenclature of this chapter, the relevant information is collated as follows:

Wheel Head of Surface Grinder

Natural frequency f_c = 38 Hz
Damping ratio ζ_c = 0.1
Maximum displacement d_r = 3 μm (Fig. 9.2*b*)

Vibration from Forging Hammer

Frequency f_s = 18 Hz
Damping ratio ζ_s = 0.1
Displacement d_s = 40 μm (Fig. 9.2*a*)
Critical element frequency ratio $\rho_c = f_c/f_s$ = 38/18 = 2.1
Transmissibility ratio required $T_R = d_r/d_s$ = 3/40 ≈ 0.07

Vibration from Diesel Set

Frequency f_s = 16.7 Hz
Displacement d_s = 20 μm rms = $2\sqrt{2}(20)$ = 57 μm p-p (see Chapter 2, Periodic Vibration)
Critical element frequency ratio $\rho_c = f_c/f_s$ = 38/16.7 = 2.3
Transmissibility ratio required $T_R = d_r/d_s$ = 3/57 ≈ 0.05

Helical Spring Isolators.   We make use of the design data for undamped mountings (Fig. 9.12), the relevant part of which is reproduced for convenience of reference in Fig. 9.21. The critical element frequency ratio and the required transmissibility ratio for the forging hammer are different from those for the Diesel set, so we consider them separately.

Forging Hammer.   Since we have a specific value of the site damping ratio ζ_s = 0.1, we can use the T_R contours for this particular excitation. Referring to Fig. 9.21, the ρ_c = 2.1 line intersects the T_R = 0.03 and 0.1 contours for ζ_s = 0.1 at points *A* and *B*. The corresponding values of mounting frequency ratio are read as ρ_m = 0.25, 0.44, respectively. The value of ρ_m for the required T_R = 0.07 is found by interpolation to be ρ_m = 0.36 (see Table 9.2, line 1). The required mounting natural frequency is then evaluated as $f_m = \rho_m f_s$ = 0.36 × 18 = 6.5 Hz.

If there were some uncertainty about the site damping ratio, the design could be based on the left boundary of the T_R envelope (points *C* and *D*, Fig. 9.21) and the required mounting natural frequency found to be 5.9 Hz as shown in Table 9.2, line 2.

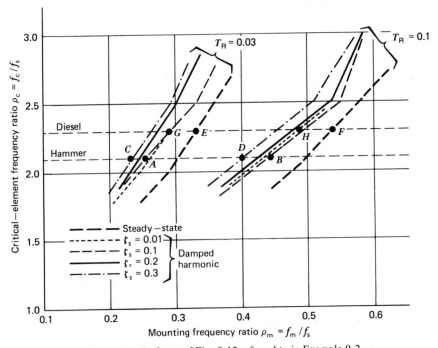

Figure 9.21.   Detail of part of Fig. 9.12 referred to in Example 9.2.

Diesel Set.   In a similar way, we find the mounting natural frequency required to attain the necessary $T_R = 0.05$ in relation to the steady-state vibration from the Diesel set. As shown in Table 9.2, line 3, the required mounting frequency ratio is found, by interpolation between points E and F on the $\rho_c = 2.3$ line in Fig. 9.21, to be $\rho_m = 0.39$. Therefore, the required natural frequency is $f_m = \rho_m f_s = 0.39 \times 16.7 = 6.5$ Hz.

During the short period immediately after the Diesel set "cuts in" the response of the wheel head relative to the workpiece is increased as a result of the starting transients of the mounting and the wheel head. The mounting natural frequency necessary to achieve the required degree of vibration isolation ($T_R = 0.05$) can be found by using the T_R contours marked $\zeta_s = 0.01$, which take these starting transients into account. The required mounting frequency ratio is found, by interpolation between points G and H (Fig. 9.21), to be $\rho_m = 0.35$ (Table 9.2, line 4). Therefore, the required mounting natural frequency is $f_m = \rho_m f_s = 0.35 \times 16.7 = 5.8$ Hz.

This is a conservative value in that the starting of the Diesel set cannot cause a sinusoidal vibration to begin instantaneously as is assumed in the derivation of the design data relating to the response including the starting transients; hence, the actual system response would be less than the theoretical.

Table 9.2. Undamped Mounting Natural Frequency Required Under Various Conditions

Condition (Fig. 9.12)	Mounting Frequency Ratio ρ_m			Mounting Natural Frequency f_m (Hz)
Forging Hammer ($\rho_c = 2.1$)	for T_R value:			
	0.03	0.1	0.07	
1. Damped harmonic	0.25	0.44	0.36	6.5
$\zeta_s = 0.1$	$(A)^a$	(B)		
2. Left boundary of T_R envelope	0.23	0.40	0.33	5.9
	(C)	(D)		
Diesel Set ($\rho_c = 2.3$)	For T_R value:			
	0.03	0.1	0.05	
3. Steady state	0.33	0.54	0.39	6.5
	(E)	(F)		
4. Damped harmonic	0.29	0.49	0.35	5.8
$\zeta_s = 0.01$	(G)	(H)		
Forging Hammer and Diesel Set ($\rho_c = 2.1$)	For T_R value:			
	0.03			
5. Left boundary of T_R envelope	0.23			4.1

^{a} Points A through H shown on Fig. 9.21.

However, this may be wholly or partly offset by the fact that the design data, although taking into account the starting transients of the mounting and the critical element, do not allow for a starting transient of the site, which would increase the site displacement d_s and hence the system response for a short period after the Diesel set cuts in.

Forging Hammer and Diesel Set. If a hammer blow occurs as the Diesel set "cuts in" the vibration at the site of the surface grinder is a combination of the transient caused by the hammer and the sinusoidal vibration from the Diesel set. The resulting site vibration depends on the relative phase of these two events. For the purpose in hand we can assume that the predominant frequency is $f_s = 18$ Hz; hence, the critical element frequency ratio is $\rho_c = 38/18 = 2.1$. The maximum displacement resulting from superposition of the two separate contributions (40 μm, 57 μm) cannot exceed 97 μm p-p. The transmissibility ratio required to keep d_r below 3 μm is $T_R = 3/97 \approx 0.03$, which is assured if the mounting frequency ratio has the value $\rho_m = 0.23$, corresponding to point C in Fig. 9.21 (see Table 9.2, line 5). Therefore,

$f_m = \rho_m f_s = 0.23 \times 18 = 4.1$ Hz. A helical spring mounting with a vertical natural frequency of about 4 Hz is indicated.

Rubber Isolators.   For rubber isolators, which normally have damping ratio less than 0.2, we conservatively use the design data in Fig. 9.13 to derive the required mounting natural frequency in the same manner as for the spring isolators. The results, given in Table 9.3, show that the natural frequency required to isolate the surface grinder from one or other source is about 4 Hz. This is below the range attainable with rubber isolators. If the damping ratio is of the order $\zeta_m = 0.1$, the required natural frequency lies between the values of 4 Hz ($\zeta_m = 0.2$) and 6 Hz ($\zeta_m = 0$), which is still below the range of rubber isolators.

To achieve the $T_R = 0.03$ which is necessary with the two sources acting together, the required natural frequency is 1.6 Hz, which is well below the range of rubber isolators (Table 9.3, line 5).

For the purpose of this example, only helical springs and rubber-in-shear isolators have been considered. The required quality of vibration isolation is attainable with a helical spring mounting designed to have a vertical natural frequency of 4 Hz. As this involves a static deflection of about 15 mm (Fig. 7.6), on springs located at floor level, the installation will have low natural

Table 9.3.   Damped Mounting Natural Frequency Required under Various Conditions

Condition (Fig. 9.13)	Mounting Frequency Ratio ρ_m			Mounting Natural Frequency f_m (Hz)
Forging Hammer ($\rho_c = 2.1$)	For T_R value:			
	0.03	0.1	0.07	
1. Damped harmonic	0.13	0.34	0.25	4.5
$\zeta_s = 0.1$				
2. Left boundary of T_R envelope	(0.09)	0.27	0.19	3.4
Diesel Set ($\rho_c = 2.3$)	For T_R value:			
	0.03	0.1	0.05	
3. Steady state	0.25	0.51	0.32	5.3
4. Damped harmonic	0.17	0.41	0.24	4.0
$\zeta_s = 0.01$				
Forging Hammer and Diesel Set ($\rho_c = 2.1$)	For T_R value:			
	0.03			
5. Left boundary of T_R envelope	(0.09)			1.6

frequencies in rocking modes. Therefore, some form of snubbing or limit stops may be required to prevent excessive horizontal movement resulting from normal operations and worktable reversals.

Rubber-in-shear isolators are unsuitable because the required natural frequency is below the range attainable with commercially available isolators of this type.

CONCLUDING REMARKS

In this chapter we have seen how the critical element can be involved as a factor in the design of a seismic mounting for vibration-sensitive equipment. Commonly the critical element is either ignored or assumed to have a relatively high natural frequency and hence to require no consideration in relation to the design of the mounting. This may result in an unsatisfactory installation if the critical element natural frequency happens to be low enough to be in the vicinity of a mounting natural frequency, and/or a component frequency of the site vibration.

To cover this possibility, the designer of the mounting should determine the characteristics of the critical element as well as those of the the site vibration, and use these in conjunction with the data presented in this chapter to determine the required mounting characteristics.

In the interests of practical application we have minimized the input data required, by presenting a conservative (worst case) approach which does not require specific values of the damping ratio and effective mass of the critical element, or of the excitation characteristics other than frequency, provided that these quantities are within the ranges covered by the calculations on which the design data are based.

Although the application of design data based on a two-mass model represents an advance on the use of the one-mass model, it still represents a high degree of idealization of an actual installation, as shown in Chapter 6 where the limitations of the two-mass model are discussed. Since the relevance and validity of the two-mass design model can be proved only in terms of the behavior of the actual installation, the designer should take any opportunity to make experimental observations of the response of the completed installation, and compare it with that predicted for the design model.

In the light of the data presented here, obviously vibration-sensitive equipment should be designed to have critical element natural frequencies high enough to be above the normal ranges of site vibration frequency and mounting natural frequency, and hence preferably above 50 Hz. The supplier of the equipment should make available to the user information about the natural frequencies of the critical elements, particularly if they are below say 30 Hz, as in some large machine tools. Designers of sensitive equipment may use these data to determine the advantage to be gained, in terms of reduced vibration sensitivity, by redesign to increase the natural frequency and damping of critical elements of the equipment.

The designers of sensitive equipment, with their intimate knowledge of the structural details of the equipment, and of the permissible relative displacement of critical elements in terms of functional requirements and specified performance, can use the data and methods given in this chapter to determine the seismic mounting characteristics they should recommend for installation of their equipment under given conditions of site vibration.

REFERENCES

Barkan, D. D. (1962). *Dynamics of Bases and Foundations*, McGraw-Hill, New York.

Biggs, John M. (1964). *Introduction to Structural Dynamics*, McGraw-Hill, New York.

Crede, Charles E. (1951). *Vibration and Shock Isolation*, Wiley, New York.

Den Hartog, J. P. (1956). *Mechanical Vibrations*, 4th ed., McGraw-Hill, New York.

Eshleman, Ronald, L. (1972). *Shock and Vibration Technology with Applications to Electrical Systems, A Survey*. NASA SP-5100, Technology Utilization Office, NASA, Washington, D.C.

Fischer, Edward, G. (1976). Theory of Equipment Design, in *Shock and Vibration Handbook*, 2nd ed., Chap. 42, C. M. Harris and C. E. Crede, Eds., McGraw-Hill, New York.

Fischer, Edward, G., and H. M. Forkois (1976). Practice of Equipment Design, in *Shock and Vibration Handbook*, 2nd ed., Chap. 43, C. M. Harris and C. E. Crede, Eds., McGraw-Hill, New York.

Gray, R. A., G. H. Cabaniss, F. A. Crowley, H. A. Ossing, T. S. Rhoades and L. B. Thompson (1972). *Earth Motions and their Effects on Inertial Instrument Performance*, United States Air Force Cambridge Research Labs., Air Force Surveys in Geophysics No. 239.

Harris, Cyril M., and Charles E. Crede (1976). *Shock and Vibration Handbook*, 2nd ed., McGraw-Hill, New York.

Hatae, M. T. (1976). Packaging Design, in *Shock and Vibration Handbook*, 2nd ed., Chap. 41, C. M. Harris and C. E. Crede, Eds., McGraw-Hill, New York.

Kaminskaya, V. V. (1964a). Investigating the Sensitivity of Machine Tools to Floor Vibrations, *Machines Tooling*, **35**(1), 3–11.

Kaminskaya, V. V. (1964b). Calculations of Machine Tool Sensitivity to Floor Vibrations, *Machines Tooling*, **35**(3), 33–37.

Kaminskaya, V. V., and E. I. Rivin (1964). Isolating Precision Machine Tools from Vibration, *Machines Tooling*, **35**(11), 7–15.

Kaminskaya, V. V. (1966). Shock Load Vibrations in Machine Tools, *Machines Tooling*, **37**(12), 3–14.

Kelly, James M., and Daniel E. Chitty (1980). Control of Seismic Response of Piping Systems and Components in Power Plants by Base Isolation, *Eng. Struct.*, **2**(3), 187–198.

Macinante, J. A., and J. K. Walter (1973). The Isolation of Machine Tools from Site Vibration, *Mech. Chem. Eng. Trans. Inst. Eng. Aust.*, **MC 9**, 1–9.

Macinante, J. A., and H. Simmons (1975). Vibration Isolating Mountings for Sensitive Equipment— A More Realistic Design Basis, *Mech. Chem. Eng. Trans. Inst. Eng. Aust.*, **MC 11**, 22–32.

Macinante, J. A. (1976). Vibration Isolating Mountings for Sensitive Equipment—New Design Criteria, *Shock Vibration Digest*, **8**(7), 3–24.

Mindlin, R. D. (1945). Dynamics of Package Cushioning, *Bell Syst. Tech. J.*, **24**, 353–461.

Morrow, C. T. (1963). *Shock and Vibration Engineering*, Wiley, New York.

Mustin, G. S. (1968). *Theory and Practice of Cushion Design*, Shock and Vibration Monograph SVM-2, U.S. Naval Research Labs, Shock and Vibration Information Center.

O'Rourke, Michael J., and George Kountouris (1979). Floor Response Spectra for Equipment, *ASCE Proc. Eng. Mech. Div. J.*, **105**(EM5), 907–911.

Rivin, E. I. (1979a), Vibration Isolation of Precision Machinery, *Sound Vibration*, **13**(8), 18–23.

Rivin, E. I. (1979b). Principles and Criteria of Vibration Isolation of Machinery, *ASME Trans. Mech. Design. J.*, **101**(4) 682–692.

Skinner, D. W., and J. L. Zable (1978). The Business Machine Vibration Environment, *Environ. Sci. J.*, **21**(5), 16–21.

Snowdon, J. C. (1968). *Vibration and Shock in Damped Mechanical Systems*, Wiley, New York.

Steinberg, Dave S. (1973). *Vibration Analysis for Electronic Equipment*, Wiley-Interscience, New York.

Waldersee, J. (1960). Response of a Seismically-Mounted Machine Tool to a Transient Site Vibration, *Aust. J. Appl. Sci.*, **11**, 250–260.

Whitehead, T. N. (1934). *The Design and Use of Instruments and Accurate Mechanism*, Macmillan, New York.

Appendix

Table 8.1.   Case Numbers for Machine Installations

Mounting Damping Ratio ζ_m	Mass Ratio R_m	Floor Damping Ratio ζ_f		
		0	0.1	0.2
0	0.1	Case 1	Case 4	Case 7
	0.5	2	5	8
	1.0	3	6	9
0.1	0.1	Case 10	Case 13	Case 16
	0.5	11	14	17
	1.0	12	15	18
0.2	0.1	Case 19	Case 22	Case 25
	0.5	20	23	26
	1.0	21	24	27

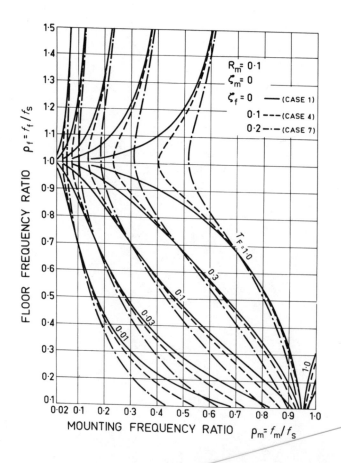

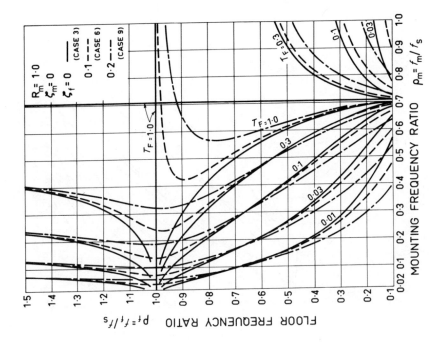

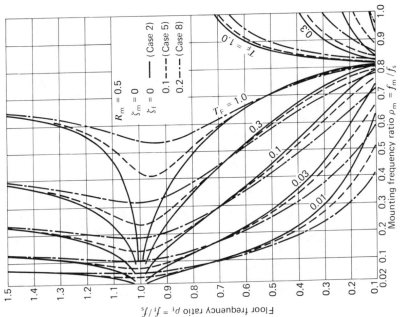

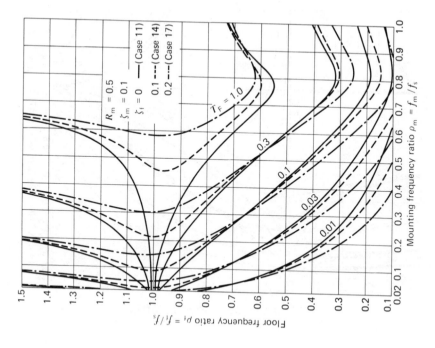

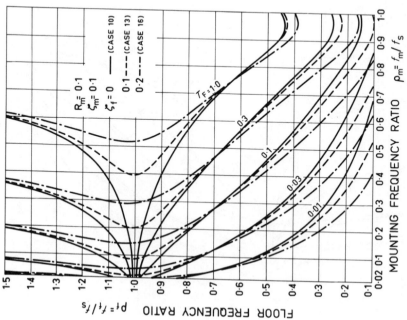

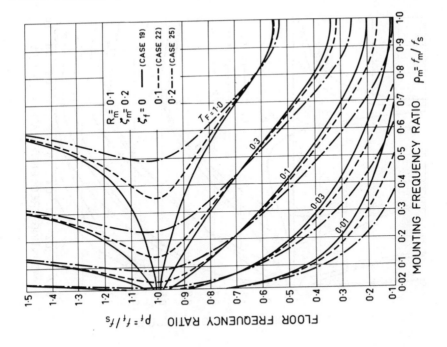

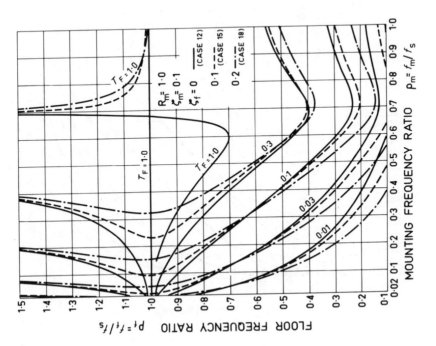

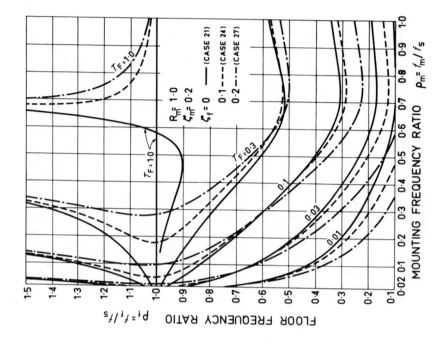

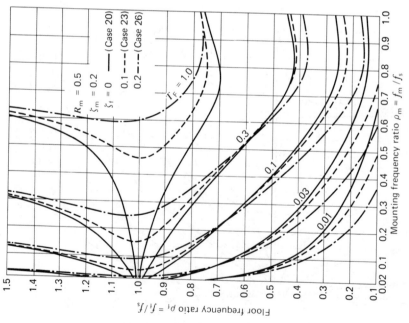

Index